SCIENCE FAIR PROJECTS

PHYSICS

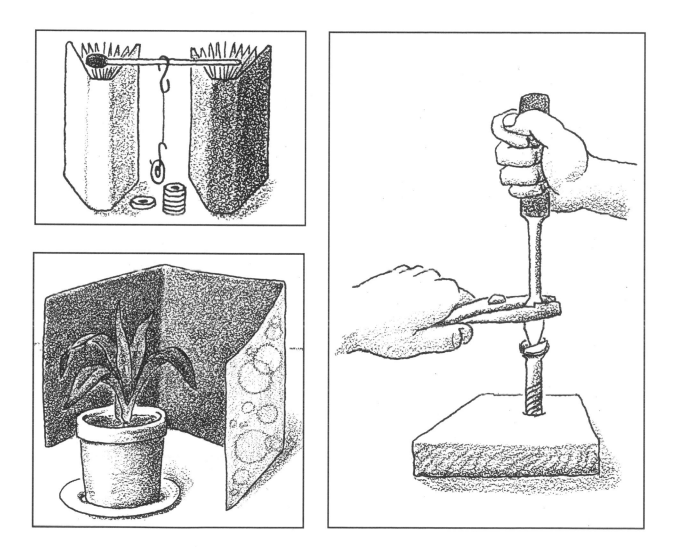

Bob Bonnet & Dan Keen

Illustrated by Frances Zweifel

Sterling Publishing Co., Inc.
New York

Edited by Claire Bazinet

Library of Congress Cataloging-in-Publication Data

Bonnet, Robert L.
    Science fair projects : physics / Bob Bonnet & Dan Keen; illustrated
by Frances Zweifel.
        p.   cm.
    Includes index.
    Summary: Presents projects and experiments that use easy-to-find materials to
explore the world of physics, covering such topics as temperature, energy flow,
acceleration, sound, pendulums, momentum, magnetism, and solar heat.
    ISBN 0-8069-0707-X
    1. Physics--Experiments Juvenile literature. 2. Science projects Juvenile
literature. [1. Physics--Experiments. 2. Experiments. 3. Science projects.]
I. Keen, Dan. II. Zweifel, Frances W., ill. III. Title. IV. Title: Physics.
QC26.B66   1999
530'¢.078--dc21   99-33938
                                                                          CIP

10 9 8 7 6 5 4 3 2 1

Published by Sterling Publishing Company, Inc.
387 Park Avenue South, New York, N.Y. 10016
© 1999 by Bob Bonnet and Dan Keen
Distributed in Canada by Sterling Publishing
℅ Canadian Manda Group, One Atlantic Avenue, Suite 105
Toronto, Ontario, Canada M6K 3E7
Distributed in Great Britain and Europe by Chris Lloyd
463 Ashley Road, Parkstone, Poole, Dorset, BH14 0AX, England
Distributed in Australia by Capricorn Link (Australia) Pty Ltd.
P.O. Box 6651, Baulkham Hills, Business Centre, NSW 2153, Australia
Manufactured in the United States of America

Sterling ISBN 0-8069-0707-X

CONTENTS

Introduction

Welcome to the fascinating world of physics! This book explores projects in the field of physics. Physics is the science of investigation that tells us the "how" and "why" about nonliving objects. It explains how a refrigerator keeps things cold, why letting the air out of a balloon causes it to fly wildly around the room, and what makes a walkie-talkie work. It tells us why we see a bolt of lightning before we hear its rumbling thunder. Physics helps us unlock the secrets of the physical world around us.

Subjects in physics are numerous, and they include: light, sound, heat, simple machines, forces, magnetism, gravity, friction, acceleration, momentum, time, space, fluidics, pendular motion, wave motion, kinetic and potential energy, work, friction, pressure, weight, conduction, the state of matter (solid, liquid, gas, plasma), electricity, radiation, and many more. Physics is one of the most interesting and motivating science topics.

It is important to understand the laws of physics because so many of its principles are found in other science disciplines, such as astronomy, geology, mathematics, health, engineering, electronics, chemistry, aviation, optics, and even the arts. For instance, meteorology, the study of weather, involves many principles that are explained by physics: convection, evaporation, condensation, temperature, precipitation, tidal action by the forces of gravity, temperature, and erosion. Many of these fields (electronics and structural engineering, for example) are really specialized branches of physics.

Physics affects our daily lives. Its principles are at work when we ride a bicycle, wear a pair of glasses, play a computer game, operate a vacuum cleaner, turn on a bedside light, play a music CD, or call a friend on the telephone. Physics is at work all around us all of the time.

Science Fair Projects

Because safety is and must always be the first consideration, we recommend that ALL activities be done under adult supervision. Even seemingly harmless objects can become a hazard under certain circumstances. If you cannot do a project safely, then don't do it!

Respect for life should be fundamental. Your project cannot be inhumane to animals. Disruption of natural processes should also not occur thoughtlessly and unnecessarily. Interference with ecological systems should always be avoided. The rules of ethics must be followed also. Consider any moral questions involved in a project and conduct yourself properly, dealing honestly with people involved as you complete the project.

Science is the process of finding out. "The scientific method" is a procedure used by scientists and in science fairs that consists of several steps: identifying a problem or purpose, stating a hypothesis, setting up an experiment to collect information, recording the results, and coming to a conclusion as to whether or not the hypothesis is correct.

A science project starts by identifying a problem, asking a question, or stating a purpose. The statement of the problem defines the boundaries of the investigation. For example, air pollution is a problem, but you must set the limits of your project. It is unlikely you have access to an electron microscope, so an air-pollution project could not check for pollen in the air. This project, however, might be limited to the accumulation of dust and other visible materials.

Once the problem is defined, a hypothesis (an educated guess about the results) must be formed. You might hypothesize that there is more dust in a room that has thick carpeting than in a room that has hardwood or linoleum flooring.

Often, a hypothesis can be stated in more than one way. For example, a project to gather data for using rocks to store and release heat during the night in a solar-heated home might test to see if a single large rock or many smaller rocks will give off stored heat for a longer period of time. This could be stated in two ways: Hypothesize that one large rock will give off stored heat for a longer period of time than an equal mass of smaller rocks. Or, you could state the opposite: Hypothesize that smaller rocks will give off stored heat for a longer period of time compared to one large rock of equal mass. It does not matter which way the hypothesis is stated, nor does it matter which one is correct. The hypothesis doesn't have to be proven correct in order for the project to be a success; it is successful if facts are gathered and knowledge is gained.

Set up an experiment to test your hypothesis. You will need to list materials, define the variables, constants, and assumptions, and document your procedure. Finally, from the results collected, come to a conclusion as to whether or not the hypothesis is correct.

When choosing a science project, try to pick a topic that is interesting to you that you would like to work on. Then all of your research and study time is spent on a subject you enjoy!

For presentation at a science fair, consider early on how you can demonstrate your project. Remember, you may not be able to control certain conditions in a gym or a hall. Decide how to display the project's steps and outcome, and keep a log or journal of how you got your results and came to your conclusion (photographs or even a video). Something hands-on or interactive often adds interest to a project display. As a fair goer, what would *your* hands be itching to do? Now is the time to pass on some of that enjoyment.

Bob Bonnet & Dan Keen

Project 1

Magnetic Water

The effect of water on magnetism

Purpose Does water affect a magnetic field?

Overview Sound waves go through both water and air. In fact, they travel farther and faster in water than they do in air. How about magnetism? Does it go through water, too?

Hypothesis Water has no effect on magnetism.

Procedure Ask an adult with a scissors to cut off the rounded top part of a 2-liter bottle. Place a metal paper clip in the bottom of the bottle.

You need
• magnet
• 2-liter plastic soda bottle
• masking tape
• metal paper clip
• 2 pencils
• string
• water
• scissors
• an adult

Wrap a strip of masking tape around one end of a six-sided, not round, pencil and then number the sides. Write "1" on the tape on one side, then turn the pencil and write "2," and so on. Tie a piece of string to the middle of the pencil and secure it with a piece of masking tape. Tie the other end of the string to a magnet. Turn the pencil, wrapping the string around it, and set it over the top of the plastic bottle. Slowly, lower the magnet into the bottle. When the magnet is close enough and captures the paper clip, stop! Notice the number on the side of the pencil.

Carefully, lift the magnet straight up without turning the pencil. Remove the paper clip, and lay it back in the bottle in the exact same spot. Fill the bottle half full with water, then slowly lower the magnet into the bottle. *Be sure not to turn the pencil*, so that the string length is not changed. The string length, the position of the paper clip, and the distance from the magnet to the paper clip are CONSTANTS. The VARIABLE is the substance between the magnet and the paper clip: air and water.

Does the magnet still attract the paper clip? If so, does it do so from about the same distance above it as it did when the bottle was filled with air instead of water?

Results & Conclusion Write down the result of your experiment. Come to a conclusion as to whether or not your hypothesis was correct.

Something more Now test magnetism using salt water, sugar water, or ice water.

Project 2

A Swinging Good Time

Pendulum motion

Purpose   The properties of pendulums will be investigated.

Overview   When a weight is hung by a wire or a string that is tied to a fixed point (a point that doesn't move), it is called a pendulum. If the weight is pulled to one side and then released to fall freely, it will swing back and forth. Gravity pulls it down, then momentum keeps it moving past the "at rest" hanging point. Eventually, the weight will stop swinging back and forth because friction with the air will slow it down. (Pendulums have been used since 1657 in clocks, because of the regularity of the swinging motion.)

> **You need**
> • 2 chairs
> • string or strong thread
> • 5 identical large metal washers (for weights)
> • hardbound book
> • long stick or pole
> • scissors

Hypothesis   Hypothesize that when the weight (mass) tied to a string is greater than one tied to another string of the same length, the heavier weight will swing longer.

Procedure   The CONSTANT in this project is the length of the string. The VARIABLE is the mass (or weight) at the end of the pendulum string.

Place two chairs back to back and a short distance apart. Lay a long measuring stick or pole across the tops of both chairs. Tie two pieces of string onto the stick some distance apart so the hanging strings almost touch the floor. Cut the strings to an equal length an inch or two (2–4 cm) from the floor. At the end of one string, tie four large metal washers. At the end of the other string, tie one large metal washer, making sure

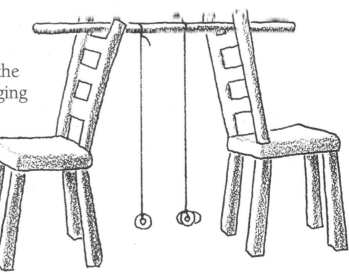

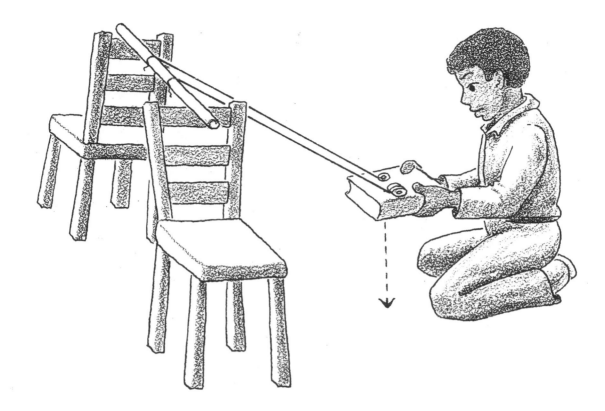

that the bottom of the washer is at an equal height from the ground as the group of four washers. In starting the pendulums swinging, you must make sure they are both released at *exactly* the same time. To do this, let the washers rest on a hardbound book and lift and pull them both, on the book, to one side of the chairs, perhaps to seat level height. Keep the two hanging strings taut. *Drop* the book down and both pendulums will begin swinging at exactly the same time. What happens then? Do they both swing at the same rate? Does the pendulum that has four washers swing four times longer than the pendulum that has only one?

Results & Conclusion   Write down the results of your experiment. Come to a conclusion as to whether or not your hypothesis was correct.

Something more

1. How does the length of the string affect the pendulum's swing? If the weights are the same but one pendulum's string is twice the length of the other, will it swing twice as long? Use the same chair set-up as above, but take the string that had four washers on it, cut it in half, and tie just one washer to it. Start them swinging at the same time. (You will have to hold one in each hand and let go at the same time as best you can, since you can't get them started together by letting them slide off a book as we did before.)

2. Think of other questions about pendulums that you can investigate and use your chair set-up to find the answers. For example, if both strings are the same length and both weights are the same, but one pendulum is pulled back farther/higher when they are set to swinging, will the one pulled back farther swing longer?

Project 3

Man on a Tight Rope

Wave motion

Purpose   Show that energy can travel along a string and do work at the other end.

Overview   Energy can travel in the form of a wave. An uncrested wave in the ocean is energy in motion. The water molecules do not travel along with the wave. That is why a boat will bob up and down when a wave goes by but does not move sideways. Surfers ride the energy of a wave, but not the actual moving water. The water only moves in a circle, but the wave energy travels forward. If you tie one end of a rope to a fixed object, such as a fence post, pull the rope tight, and then give your end a quick snap up and down, with a fast wrist movement, you will see a wave-like motion travel along the rope to the fixed end. That is wave energy moving along the rope, but any spot on the rope only moves up and down. You can see this easily by making colored markings along the rope and watching them bob up and down. Now, let's track the energy.

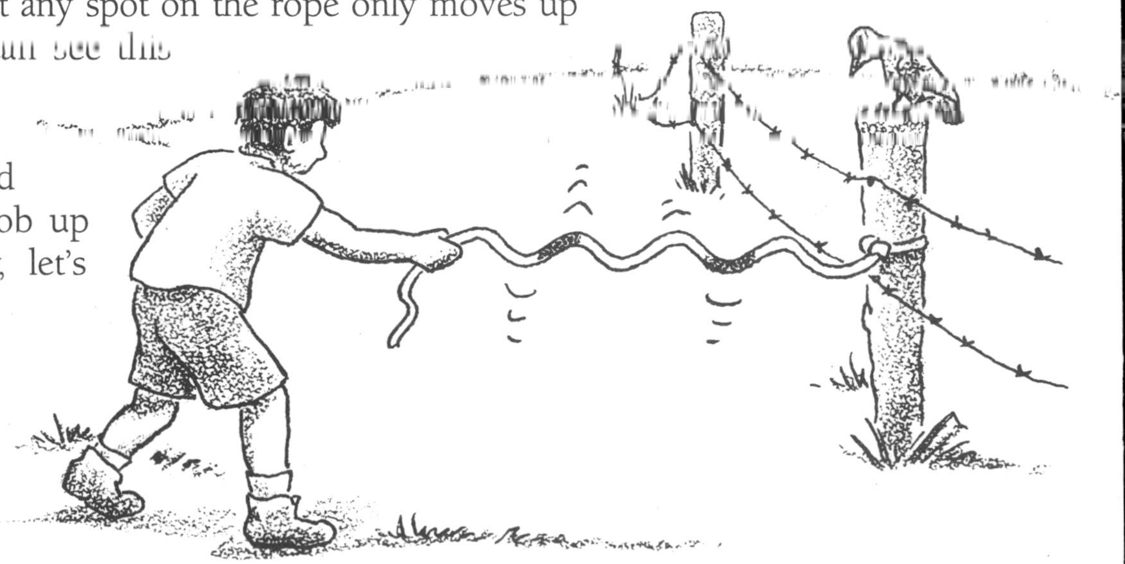

You need
- length of string
- 2 chairs
- scissors
- small piece of paper
- pencil

10

Procedure   Set two chairs about 4 feet (125 cm) apart. Tie a string tightly from the back of one to the other. Cut a small square piece of paper and fold over one-quarter of it to act as a hook. Hang the piece of paper onto the string near one end by the fold in the paper. Near the other end, hit the string hard with a pencil. You have put energy into the string by making it move up and down. Does that energy travel along the string and flip the piece of paper off the string at the other end?

Results & Conclusion   Write down the results of your experiment. Come to a conclusion as to whether or not your hypothesis was correct.

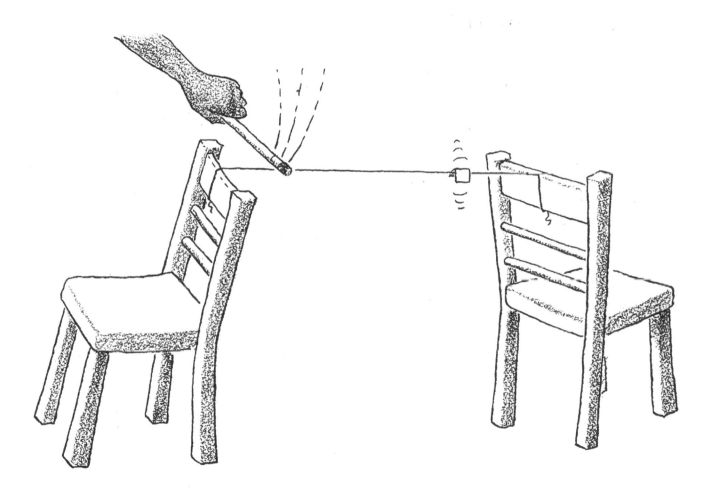

Something more   Make a colored mark in the middle of a long piece of rope, such as a jump rope or clothes line. Hold one end of the rope and have a friend hold the other. Wave your arm with the rope up and then down, and have your friend do the same but in the opposite direction: when you go up, he goes down. Is it possible to synchronize, or time, both your movements so that the mark in the middle of the rope will not move up or down?

Project 4

Rub the Right Way

Friction and surfaces

Purpose Compare the friction on a dry surface to one coated with oil.

Overview Friction is the resistance to motion when two things rub together. Friction is often undesirable. It makes machines less efficient where moving parts come in contact with each other. But there are times when friction is helpful. On the road, it's the friction between a car's tires and the road's surface that allows a driver to keep control of the car. If a road becomes covered with water, snow, ice, or spilled oil, the car becomes harder to steer and to stop. This is especially true on a hill.

You need
- 2 pieces of wood (about 2 feet long (60 cm)
- 2 small plastic butter tubs with lids
- sand
- vegetable oil
- an old rag
- several books
- ruler
- protractor

Hypothesis Hypothesize that if friction becomes less, an object on a slope will need less of an angle for gravity to overcome friction.

Procedure Make a ramp (the slope) using a piece of wood about 2 feet (60 cm) long and 3 to 4 inches (7–10 cm) wide (a 2-by-4 board works well). To raise one end of the ramp, place several books under one end.

Using an old rag, wipe some vegetable oil onto the board, coating and completely covering the surface. This represents spilled oil on a roadway.

Fill two empty plastic butter tubs with an equal amount of sand, and close the lids.

Place one of the filled tubs in the center of the board. By adding more books or pushing them a little farther under the ramp, slowly make the ramp steeper until gravity overcomes the friction between the surfaces and the tub moves. When this happens, stand a ruler alongside the highest point of the ramp. Measure and write down the height of the ramp at that point. Then, using a protractor at the low end, measure the angle, or slope, of the ramp from the table or floor.

Using books and another board, make a ramp with the same slope as the first ramp. Place the tub in the middle of the board. This time, the tub does not move. Slowly raise the slope of the ramp by adding books until the tub finally moves. Measure the angle of the ramp, using the protractor, and see how much steeper it is compared to the first ramp. The weight here is now the CONSTANT, and the surface friction is the VARIABLE.

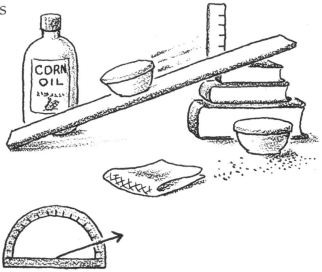

What other places can you think of where friction is desirable? Think about walking on patches of ice outside, a newly waxed kitchen floor, or the tile floor in the bathroom when you step out of the shower.

Results & Conclusion   Write down the results of your experiment. Come to a conclusion about your hypothesis.

Something more   Instead of comparing an oil-covered surface to a dry surface, compare a dry surface to one that is covered with ice. Place a piece of wood under the faucet in a sink and run water on it. Then put the wet piece of wood in the freezer and leave it there until the water has turned to ice. Again, find the angle where gravity overcomes resistance and the sand-filled tub moves. Do you think driving a car on an ice-covered road is more dangerous than when the road is dry? Besides driving more slowly, what do people do to help make driving on snow and ice safer?

Project 5

The Mighty Mo

Momentum: a product of force times mass

Purpose Can momentum be increased by increasing the mass of the moving object or by increasing its speed?

Overview Momentum is the force with which an object is moving. Objects in motion tend to stay in motion. When you start a ball rolling, it keeps rolling until friction with the surface on which it is rolling and air resistance slows it down. Momentum is a factor of mass and velocity. Mass is a measure of how much "stuff" an object is made of and velocity is how fast an

object is moving and in what direction. If either the mass or the velocity is increased, the momentum will be increased, and the moving object will have more force.

Hypothesis Two hypotheses can be stated for this experiment. As the speed of an object increases (speed is affected by the drop point being raised), the momentum increases (as measured by the depth of the hole in soft material). As the mass (weight) of an object increases, the momentum increases.

Procedure We can measure the momentum of a falling ball by dropping it into soft dough. Make a batch of dough by mixing water, salt, and flour. The dough must be thick enough to keep a golf ball from going through a 2-inch-thick (5 cm) batch of it when dropped from a height of about 4 feet (120 cm), but soft enough so that a ping pong ball dropped from the same height will make a small impression.

Cut a piece of cardboard about a foot (30 cm) square. Cover the cardboard with the layer of dough.

Drop a golf ball into the dough from a height of 12 inches (30 cm). The impact will make a depression in the dough. Now hold the golf ball in your hand and raise your arm as high as you can. Drop the ball into another spot in the dough.

Compare the two depressions. Did the ball have more momentum and more force

when it was moving faster? Measure the volume of each depression by using an eye dropper to fill each depression with water. Count the number of drops each depression takes to fill it.

In the above experiment, the mass was kept as CONSTANT, but the velocity was increased (the VARIABLE). Now let's keep the velocity CONSTANT and increase the mass (the VARIABLE).

Hold a ping pong ball in your hand and raise your arm as high as you can. Drop the ball into a clear spot in the dough. A ping pong ball and a golf ball are about the same size, but the golf ball has more mass. They were both dropped from the same height, so they were traveling at the same velocity when they hit the dough. Did the ball with more mass have greater momentum and hit the dough with greater force?

Results & Conclusion Write down the results of your experiment. Come to a conclusion about the two hypotheses.

Something more If you have two different velocities (by dropping two balls, each from a different height), can you adjust the mass of one of the balls to make the momentum equal? A small hole can be cut in a ping pong ball to allow different quantities of water added to it, making it heavier (increasing its mass).

Project 6

Hot Rocks

Heat transfer from one medium to another

Purpose   Is there a good way to store solar heat, and release it slowly over time?

Overview   Did you ever touch a rock that has been baking in the sun on a warm summer day? Did it feel hot? Rocks can collect and store heat.

Scientists have been working for many years to harness energy from the sun. Solar energy is being used to heat houses. One design uses hollow roof panels so that the sun warms the air inside. A fan blows the warmed air through a pipe to the basement, which is filled with rocks. As the heated air flows over the rocks, heat is transferred from the air to the rocks, warming them. Then at night, when the collectors no longer gather solar heat, a fan blows air over the rocks, transferring their warmth back to the air. The air is sent to ducts throughout the house to warm each room.

In designing a solar-heated house like this, would it make any difference if huge rocks were used or very small ones? A big rock might have more ability to store heat, but many smaller rocks would have more surface area (they have more sides that would be exposed to the warm air). Find out if a big rock or many smaller rocks would be better at collecting heat, or if rock size doesn't seem to make much of a difference.

Hypothesis   Hypothesize that an equal mass of smaller rocks will absorb heat more quickly than one large rock.

Procedure   Have an adult help you by cutting the tops off two 2-liter plastic soda bottles, using a pair of scissors. They should be cut near the top, just at the point where the bottles start to become rounded.

Gather some rocks. One of the rocks should be just large enough to fit inside a 2-liter soda bottle, about 3 inches or 8 centimeters in diameter (across). The other rocks should be small, pebbles about the size of small coins.

Using a scale, find out how much the large rock weighs. Remove it from the scale. Then pile up smaller rocks on the scale until the same weight is reached. The rock mass will then be held CONSTANT, and the size of the rocks is the VARIABLE.

Set all the rocks on a table for an hour or two until you can be sure they are all at room temperature. Do not put them in direct sunlight.

Gather two thermometers. Before we can use them, we must be sure they are calibrated, so that we can use their readings for comparison. (We might have to adjust the reading of one thermometer to correct it so both thermometers read the same temperature.) Leave the two thermometers at room temperature for several minutes, then read the temperature on each one. If one reads higher than the other, put a small piece of masking tape on it and make a note of the difference in temperature. If it is ½ or 1 degree higher than the other, then subtract this much from its reading when comparing the temperature on it to the temperature on the other thermometer.

Have an adult fill each bottle half full of hot water from a sink. Using a thermometer, be sure the water in each bottle is the same temperature. Be careful working around the bottles of very hot water.

Place the large rock in one 2-liter plastic bottle and the smaller rocks into the other bottle. Be careful not to spash the hot water out and on you.

Put a thermometer in each bottle. After a few minutes, record the temperature on each thermometer. Every three minutes, read and record the temperature on the two thermometers. Make up a table, such as shown in the illustration, to record your data. Continue to record temperatures until they reach room temperature (about 70 degrees Fahrenheit). Remember to make an adjustment of your readings to calibrate the two thermometers.

[1] -1°		[2]
120° F		119° F
118° F		117° F
115° F		

Did the water in one bottle cool off faster than the other? If so, then the rock (or rocks) in that bottle collected heat faster.

Results & Conclusion   Write down the results of your experiment. Come to a conclusion about your hypothesis.

Something more   Which releases heat quickest, one large rock or an equal-mass grouping of smaller rocks? In solar heating for a home, it would be preferable to have heat released slowly over a long period of time, to keep a house warm all through the night until the sun came up again to add heat back into the system.

Project 7
Smaller Is Stronger
Testing tensile strength

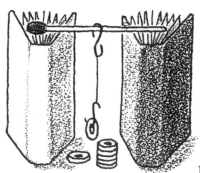

Purpose To discover if an object's strength has any relation to its length.

Overview The term tensile strength means how strong something is when it is unsupported; how much tension or pressure it can take before it breaks. Steel has great tensile strength. Is tensile strength affected by length?

You need
- large metal washers
- 2 paper clips
- string
- 2 hardbound books
- an adult
- 2 long (fireplace) safety matches
- paper and pencil

Hypothesis As an unsupported span decreases in length, it can support more weight.

Procedure Stand two hardback books upright, opening them slightly. Place them about 10 inches (25 cm) apart. *Have an adult* light and blow out long matches, made specially for fireplaces, so they are safe to use. Lay one match across the books. Bend open two metal paper clips so they form an "S," with a hook at the top and bottom of each clip.

Tie a paper clip onto each end of a short piece of string. Hang one paper clip from the middle of the match. Push the hook of the other paper clip through the hole of a large metal washer. This makes it easy to add more washers.

Add washers until the match breaks. Write down how many the match could hold.

Now repeat the experiment, but this time move the books closer together, about half the distance they were. The VARIABLE is the length of the span being stressed. Will the shorter unsupported span of the match be able to hold more weight without breaking?

Results & Conclusion Write down the results of your experiment. Come to a conclusion as to whether or not your hypothesis was correct.

Something more Can you work out (quantify) the relationship between the length of unsupported match in inches and the number of washers needed to break it?

2" = 8 washers   6" = ? washers

Project 8

Up to Speed

Acceleration in a bottle

Purpose   To show changes in rate of speed.

Overview   Acceleration is an increase in speed. Physicists define it as a measure of the rate of change of velocity over time. To accelerate means to go faster; to decelerate is to slow down.

Hypothesis   It's possible to prove that speed and acceleration are different and measurable by constructing an "accelerometer."

Procedure   Partially fill a 2-liter clear plastic soda bottle with water. Add some food coloring so that you will be able to see the water's movement better. Screw the cap on tightly. Place masking tape along one side of the bottle's circumference. Draw a scale on it (millimeters or ¼-inch increments).

Ask an adult with an automatic-shift car to take you for a short drive. (Manual-shift cars could be jerky and uneven during acceleration.) Open the car's glove compartment and use the door as a shelf. Lay your "accelerometer" on it. Fix it there with masking tape or rubber bands. On the tape, mark the water level when the car is not in motion.

When you are ready (remember to fasten your seat belt), have the driver accelerate. Observe how far up the scale the water moves. The faster the car accelerates, the steeper the slope of the water up the side of the bottle. The amount of water in the bottle has remained CONSTANT, but the acceleration of the car has VARIED.

Next, have the driver hold a steady speed, like 40 miles or kilometers per hour, on a highway. Is the water level at the same mark as it was when the car was at rest? Even though the car is traveling at quite a good speed, the acceleration is zero.

Results & Conclusion   Write down the results of your experiment. Come to a conclusion as to whether or not your hypothesis was correct.

Something more   What about deceleration, when the car is slowing down and coming to a stop. Can your accelerometer also be used to compare rates of deceleration?

Project 9

Bad Manners

Heat conduction and heat sinking

Purpose Is there a way to make something cool more quickly, like a drink that is too hot?

Overview Metal is a good "conductor" of heat. That means it makes an easy path that heat can travel along. When a metal frying pan is placed on a stove burner, the heat from the burner is conducted (carried) through the bottom of the pan and heats the food inside it.

Metal is sometimes used to cool things by conducting heat away from an object. In electronics, transistors and integrated circuits ("chips"), which are found in televisions, stereos, and computers, get hot, but heat can damage them. Often metal is made in the shape of fins and attached to transistors and integrated circuits in order to carry the heat away from them. These cooling fins, called heat sinks, help transfer the heat to the surrounding air and keep the transistors and integrated circuits cool. Sometimes a small fan is used to get rid of the heated air. Does your computer have a fan in it?

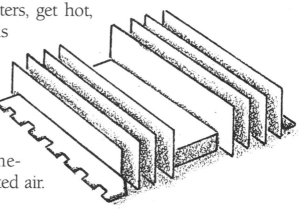

> ### You need
> * 2 identical containers (coffee mugs or tea cups)
> * hot tap water
> * 2 thermometers
> * masking tape
> * spoon
> * clock or watch
> * paper and pencil

Have you ever been served a hot cup of tea or hot chocolate that was too hot to drink and someone told you, "Leave the spoon in. It might be bad manners but it will help cool the drink faster." They are thinking that, since metal conducts heat, the spoon will carry some of the heat away from the drink. The spoon is indeed hot to the touch, so it does conduct heat away from the drink.

However, since the handle of the spoon is not designed like a heat sink, the heat in the spoon doesn't efficiently transfer to the surrounding air, so you may want to hypothesize that leaving the spoon in the hot liquid won't make a significant difference.

Hypothesis Hypothesize that leaving a spoon in a hot drink will not make any noticeable difference in its rate of cooling, reducing the temperature faster.

Procedure Gather two thermometers. Before we can use them, we must be sure they are calibrated, that is, we may need to adjust the temperature readings of one thermometer so both thermometers correctly read the same. Leave the two thermometers at room temperature for several minutes, then read the temperature on each one. If one reads higher than the other, put a small piece of masking tape on it and make a note of the difference in temperature. If it is ½ degree or 1 degree higher, then subtract this much from its readings when comparing the temperature on it to the temperature on the other thermometer.

Fill two coffee cups of equal size with equally hot tap water. (Be careful working with and around very hot water.) Place a thermometer in each cup. Put a metal spoon in one of the cups. After one minute, read the temperatures on the two thermometers and write them down. Every minute, write down the temperatures you read. Be sure to make any adjustment of your numbers to calibrate the two thermometers. Continue to make readings until the water in the two cups reaches room temperature.

Did the water in the cup with the spoon in it cool down faster, or wasn't there any noticeable difference?

Results & Conclusion

Write down the results of your experiment. Come to a conclusion as to whether or not your hypothesis was correct.

Something more
Can you find a way that will measurably cool the cup of hot water? Purchase some transistor heat sinks at your local electronics shop and affix them to the cup with rubber bands. Try using larger spoons, such as a ladle.

Project 10

Watt?

Comparing light output and power consumption

Purpose Determine if a 100-watt light bulb gives off as much light as two 50-watt bulbs.

Overview Incandescent light bulbs, the kind used in most household lamps, are rated by the amount of power they use, but the amount of light they give off is really the most important thing to know. Light bulbs are rated by the number of "watts" they use. A watt is a unit of measure of electric power, that is, how much electrical energy is used. It would take the same amount of electrical energy to light one 100-watt bulb as it does to light two 50-watt bulbs, so the cost would also be the same. But, does a 100-watt light bulb give off as much light as two 50-watt bulbs?

> **You need**
> - an adult (for safety when working with electricity or hot light bulbs)
> - camera
> - 2 lamps
> - 1 hundred-watt light bulb
> - 2 fifty-watt light bulbs
> - a room that can be made completely dark
> - an index card or stiff piece of paper
> - dark marker

Hypothesis Hypothesize that a 100-watt light bulb will give off about the same amount of light as two 50-watt light bulbs.

Procedure Place two lamps side by side on a small table, dresser, or any object that will hold them in a room that can be made dark. Before plugging the lamps into an electric outlet, screw a 50-watt bulb in each lamp. Be very careful plugging the lamps into the outlet. You can leave the lamp shades on or take them off, but whichever you do, you must do the same when you use the single 100-watt bulb later. If the lamps have different shades, then you must remove both shades.

Fold an index card or small piece of paper in half, making a "V" shape. Turn it upside down so it will stand up. On one side, write "1" with a dark marker. On the other side, write "2." Place it on a table, nightstand, or dresser on the side of the room opposite the lamps. Stand the card so that the 1 is visible.

Turn the lamps on. Stand with your back to the lamps, but be sure that your body is not blocking the light from shining directly on the card on a table or dresser. Face the card and the rest of the room to take a picture. Be sure the camera does not have a flash, or that the flash is turned off. The camera must also be one that does not have an automatic sensor for lighting. Take a picture, focusing on the index card. You may want to set the camera on a table or something to keep it still, and to ensure that the camera will still be in the same position for the next picture. The camera position and everything in the room will remain CONSTANT. The only VARIABLE will be the light bulb(s).

Turn the lamps off. Unplug one of them from the electric outlet, unscrew the 50-watt bulb, and replace it with a 100-watt bulb. Plug the lamp back in and turn it on. Turn the index card around, so the side with "2" on it is showing. Again, stand with your back to the lamp and take a picture, with the index card as the focal point.

If you have to send the film away to be developed, write down on a piece of paper that the photo with the #1 on the index card was taken with two 50-watt bulbs, and the one with #2 was taken with one 100-watt bulb. That way you won't have to remember which photo matched which lighting experiment.

Compare the two pictures. Do objects in the pictures have about the same brightness, or are there differences?

If you can borrow a light meter from a photographer or your school's science teacher, try to measure the amount of light given off by two 50-watt bulbs and compare it to the light given off from one 100-watt bulb.

Results & Conclusion   Write down the results of your experiment. Come to a conclusion about your hypothesis.

Something more

1. Even if two 50-watt bulbs give off about the same amount of light as one 100-watt bulb, do you think two 50-watt bulbs are better for lighting a room? By "better" we mean that the light is more evenly distributed and less harsh, making it easier to read, work, or play in a room when there are two lights on opposite sides of the room rather than just one real bright one.

2. Audio power (volume) is also measured in watts. Does a stereo sound louder if its two speakers are placed next to each other, or spread far apart?

Project 11

Room for Brightness

Reflected light

Purpose Show that a room is better lit when the room's walls are painted in bright colors compared to a room where the walls are dark (makes a room safer, reduces eye fatigue when reading or working, and makes the room a more cheerful, pleasant place to be).

Overview In a house, some rooms are brighter than others, not just because they have more indoor lighting or windows to let sunlight in but because the walls, ceilings, and floors are more brightly colored. In a kitchen, people need lots of light to work with food. A bright bathroom makes it a safer place. A living room, bedroom, or den, however, may have darker, more deeply colored carpeting and dark walls or paneling for a quiet feeling of richness and luxury.

Bright colors, such as white and yellow, reflect much of the light that hits them. When walls, ceilings, and floors are bright in color, more light is reflected (bounced) off of those surfaces, and the light spreads around the room.

You need
- a room with light, brightly colored walls and that can be made completely dark
- a room with dark-colored walls and that can be made completely dark
- camera (an instant camera is preferred, but not required)
- lamp
- tape measure
- an index card or stiff piece of paper
- dark marker
- a friend

Hypothesis Photographs can be used to show how a room that has bright colored walls is brighter than a similar room that is darkly colored.

Procedure Find two rooms about the same size in a house; one room that has light colored walls and another room that has dark walls. The rooms must be able to block any light coming in from outside of the room, such as car lights or street lights through a window, even the glare of a television set from another room. Do this experiment at night to reduce outside light from leaking in behind curtains or blinds.

Take an index card or stiff piece of white paper On one side, write "#1" with a dark marker. On the other side, write "#2."

In a room with light walls, place a lamp on a table, dresser, or any object that will raise it up to the height of a normal table. Place the table and lamp against a wall. Turn the lamp on. Using a tape measure, stand 4 feet in front of the lamp, with your back to it. Have a friend stand 10 feet from the lamp, 6 feet in front of you, facing you and holding the index card with the #1 side facing the camera. You are standing *between* your friend and the lighted lamp. Take a picture of your friend. (Do not use a camera that has an automatic flash or an automatic lens adjustment for light levels.)

Next, in a room with dark walls, place the same lamp on a table, dresser, or any object that will raise it to the same height as it was in the lighter colored room. Place the table and lamp against a wall. Turn the lamp on. Again, using a tape measure, stand 4 feet in front of the lamp, facing away from it. Have your friend stand 10 feet from the lamp, face you, and hold the index card with the number #2 side facing the camera. Take a picture of your friend.

If you take or send the film for developing, tell them *not* to "adjust" prints. Also, write yourself a reminder that the #1 card photo was taken in a light-colored room and the #2 one was taken in a dark-colored room. The light source and distance of your friend from the camera remain CONSTANT. The VARIABLE is the color of the walls.

Compare the two pictures. Even though the same amount of light was used in both pictures, did the picture of your friend come out darker in the room that had the darker walls?

Results & Conclusion Write down the results of your experiment. Come to a conclusion as to whether or not your hypothesis was correct.

Something more Do you think the color of the ceiling and carpet on the floor also affects how light is reflected in a room?

Project 12
Down-Range Shooter
Trajectory: curved path through the air

Purpose The purpose of this experiment is to determine the angle of trajectory that will give the greatest distance.

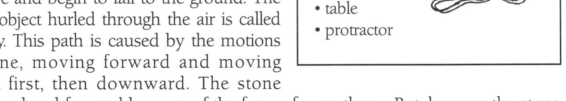

You need
- an adult
- wooden ruler
- utility knife
- rubber band
- dominos
- table
- protractor

Overview When you throw a stone a little upward and away from you, you know that it will not keep going in that direction, but will slowly curve and begin to fall to the ground. The path of an object hurled through the air is called its trajectory. This path is caused by the motions of the stone, moving forward and moving upward at first, then downward. The stone moves upward and forward because of the force of your throw. But, because the stone has weight, the Earth's gravity eventually causes it to curve and fall down.

What determines where an object will land when it is thrown or launched? The force at which it is thrown and its upward angle are both factors.

The trajectory of an object is very important to the military when they use artillery. If a cannon is to hit its target, the operator has to know the right angle to tilt the gun upward.

Hypothesis Hypothesize that the launch angle of an object will affect the distance it will travel away from the launching device.

Procedure Because of the hazard of using a sharp tool, have an adult cut a small notch at each inch or centimeter mark on a wooden ruler with a utility knife or razor.

Stretch a rubber band from one end of the ruler to one of the notches at a marking, giving it a good stretch, but not to its maximum stretch potential. Hold the ruler lengthwise with one hand near the edge of a table. With the other hand, push the rubber band out of the notch until it launches. Place a domino on the floor to mark the spot where it landed. Always keep safety in mind; do not launch the rubber band while anyone is standing in front of the ruler.

Then raise the end of the ruler by placing dominos under it. Experiment with different heights (elevation). By launching the rubber band from the same marking, the launch force is kept CONSTANT, and only the angle is VARIABLE.

Set a protractor on the table and measure the angle the ruler is elevated upward before each launch. What is the angle that shoots the rubber band the farthest distance from the ruler? What is the angle that, if the angle is further increased, the rubber band will not travel any farther (this is the optimum angle for distance)?

Note: If the rubber band gets caught on some of the other notches in the ruler as it launches, you can put tape over them, or make a horizontal cut in a straw and cover the notches.

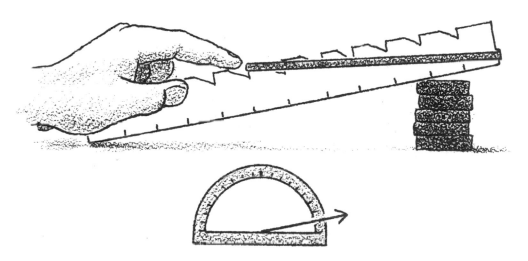

Results & Conclusion   Write down the results of your experiment. Come to a conclusion as to whether or not your hypothesis was correct.

Something more
1. The distance the rubber band will travel depends on both the launch angle and the launch force. Try launching the rubber band from different notches, which changes the force. The more the rubber band is stretched, the more force it will have when it is launched. What combination of tilt angle and launch force (marked notch) makes the rubber band travel the farthest? The highest?
2. Make a game by having a friend place a domino on the floor while you try different combinations of force and tilt to land the rubber band as close as you can to the target. Then reverse places with your friend, and see who can come closest with the least number of tries.

Project 13

Get a Handle

Wheel and axle, a simple machine

Purpose Understanding the wheel and axle concept and what a great advantage a screwdriver handle gives you because it is a "simple machine."

You need
• screwdriver with a large handle
• pair of pliers
• a friend

Overview Simple machines is the term used in physics to refer to a group of tools that make it easier to do work. Levers, inclined planes, pulleys, wedges, screws, and wheel-and-axles are examples of simple machines. With the simple machine called a wheel and axle, movement over a bigger distance gives a stronger force over a smaller distance.

Imagine trying to open a door if the door knob is missing and only the small shaft is there! The door knob is like a big wheel; the knob covers a bigger distance when it is turned but makes it easy to turn the tiny shaft. Think about steering a car without the steering wheel or closing a submarine's watertight hatch without the wheel on the door.

A wheel and axle tool may not actually have a wheel but simply a "spoke" that is rotated in a circular direction, like a wheel. An example of this type of wheel and axle is a pencil sharpener. When you turn the handle, you are covering a larger distance than the turning sharpening blades inside, but a greater force is gained at the blades. Using a socket wrench is another example, where you turn the handle over a larger distance in order to turn a nut a smaller distance, but with greater force.

Sometimes the opposite is needed—a little distance and a lot of force yields a smaller force but a greater distance. An example of this is swinging a baseball bat. At the handle, you put a lot of force into the bat and turn over a small distance, and the other end of the bat moves over a greater distance, giving it more speed.

Show the concept of wheel and axle using a screwdriver and a few friends. A screwdriver is a simple machine—a wheel and axle. Turning the bigger handle gives a bigger force at the smaller blade end.

Hypothesis No matter how much I try, I will not be able to turn a screwdriver by the blade using only my hand while a friend holds the handle.

Procedure   Have a friend hold the handle of a fairly large screwdriver. Tell your friend to try to keep it from turning. Grab the blade with one hand and try to turn the screwdriver while your friend holds the handle. Even if your friend is weaker than you, you will not be able to turn the screwdriver. Try to grasp and turn the shaft using two hands instead of one. Are you able to turn it at all?

Using a pair of pliers, now try to turn the blade (hold the pliers perpendicular to the screwdriver shaft). Is your friend still able to hold the handle and keep it from turning? What if your friend uses two hands?

The pliers are a wheel and axle too, and the pliers form a much bigger spoke of the wheel than the screwdriver handle. The longer the length of the turning spoke, the more force you get from it, but the more distance must be traveled.

Results & Conclusion   Write down the results of your experiment. Come to a conclusion as to whether or not your hypothesis was correct.

Something more   Try to turn a long thin screw into a piece of wood with your bare hands by turning the shaft. Is it impossible? Then try using a screwdriver. Is it easy to do, or easier? If the wood is very hard, try using the pliers to help turn the shaft while you press down on the screwdriver. What happens now?

Project 14

Balance the Books

First-class lever, a simple machine

Purpose How can you manage to raise something that is to heavy for you to lift, or that you want to lift using less force?

Overview Simple machines is a term used in physics to refer to a group of tools that make it easier to do work. Levers, inclined planes, pulleys, wedges, screws, and wheel-and-axles are examples of simple machines.

With the simple machine called a lever, movement over a bigger distance provides a stronger force at the other end, but moving over a smaller distance. This is done by placing a support called a fulcrum under a long shaft or board. When the fulcrum is placed closer to the load (the object to be moved), it is called a first-class lever (there are three classes of levers—see Project 15). A pry bar is an example of a first-class lever tool.

See the first-class lever drawing below. A push down on the end of the lever that extends farther out from the fulcrum will yield a gain in force at the other end. However, a much greater distance is covered in order to raise the object only a small distance.

Hypothesis It takes less effort to lift a group of heavy books with a first-class lever than to lift them by hand.

> **You need**
> - 7 heavy books (encyclopedia volumes, for example)
> - 1-by-4 piece of lumber, 4 feet (120 cm) long
> - rope
> - bathroom scale
> - clock or watch

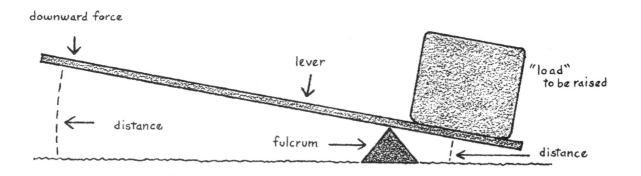

Procedure   To get an idea how hard it is to lift a heavy book against gravity, pick up a heavy book while keeping your arm straight, then hold it at arms length for one minute.

Gather seven heavy books (encyclopedia volumes, for example). Tie six of the seven books together with a piece of rope. Place them on a bathroom scale to see how much the group of books weighs.

Place the one loose book underneath the board, about one foot (30 cm) from an end. This book will be the fulcrum. Place the tied books on the board at the end nearest the fulcrum. Press down on the other end of the board.

Were you able to raise the pile of books with little effort? Even though they were only raised a short distance, it would have been much harder to lift the books the same distance straight up off the ground yourself. Remember putting them on the scale?

Results & Conclusion   Write down the results of your experiment. Come to a conclusion as to whether or not your hypothesis was correct.

Something more

1. If you have a small brother or sister, remove the pile of books and have him or her sit on the board, and use the lever to raise this load.

2. Is it harder to lift an object as you move the fulcrum farther away from the load and closer to the end where the effort (the force) is being applied?

Project 15

We Are #2

Second- and third-class levers

Purpose Learn to use a second-class lever (a simple machine) to reduce the force required to lift a heavy object.

Overview A lever is made up of a long shaft or board that rests on a support called a fulcrum. The object you are trying to move is called the load. With a lever, either force is gained and distance is lost or distance is gained and force is lost.

> **You need**
> • wheelbarrow
> • 3 plastic, one-gallon jug bottles
> • rope
> • water

A seesaw is a lever. Some seesaws allow the fulcrum to be moved from the middle balancing point. If you sit on a seesaw and the fulcrum is closer to the other end of the board, a person sitting on that end will be easier for you to lift than if the seesaw had the fulcrum in the middle. This is an example of a first-class lever (see Project 14). A first-class lever is useful for lifting heavy objects a short distance.

There are three classes of levers, first class, second class, and third class. A third-class lever is the opposite of a first-class lever. In a third-class lever, the fulcrum is close to where the force is applied. Sweeping with a broom is an example of a third-class lever. Your one arm applies a force, moving a small distance. Your other arm is the fulcrum, the point that the lever swings around. The other end of the lever has less force, but makes a big gain in the distance it moves. When you cast a fishing line into the water,

First-Class Lever   Second-Class Lever

you are using the concept of a third-class lever. Your one arm is the fulcrum and the other arm applies a lot of force over a short distance. The end of the pole has less force but moves quickly over a longer distance to throw the line out into the water.

In a second-class lever, the fulcrum is actually the point on which the lever rests, and instead of pushing down with force, you lift up (applying an upward force). Like a first-class lever, this too is useful for lifting heavy objects to a short height.

Hypothesis   It is easier to lift a heavy object by using a second-class lever, and the closer the load is to the fulcrum, the easier lifting becomes.

Procedure   First, fill three plastic one-gallon jugs with water. Water weighs a little more than 8 pounds per gallon. Try gently lifting all three jugs. Careful! They're heavy!

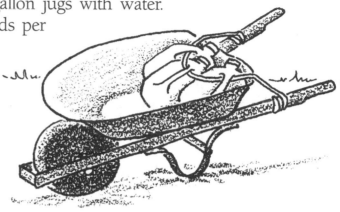

Now, place the three jugs (one at a time) in a wheelbarrow at the end near the wheelbarrow's handles. Lay them in a row and use a piece of rope to tie them together and to the handles of the wheelbarrow. This will keep them from sliding when you lift the wheelbarrow up.

Lift the wheelbarrow by its handles. You will find that the jugs of water (including some of the weight of the wheelbarrow) is easier to lift then just trying to lift the jugs by themselves.

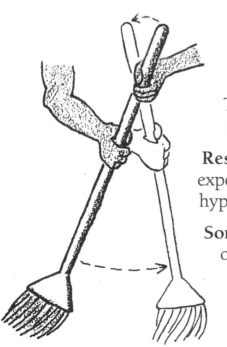

To make them even easier to lift, untie the jugs and move them to the front of the wheelbarrow. Now they are almost over the wheel, which is the fulcrum. Again, lift the wheelbarrow by its handles. We lost distance (the jugs didn't raise quite as high), but we gained force. The jugs are even easier to lift when they are closer to the fulcrum.

Results & Conclusion   Write down the results of your experiment. Come to a conclusion as to whether or not your hypothesis was correct.

Something more   Use a second-class lever to pull up on an object, rather than push up. Using a long board, place one end on the ground. Tie a piece of rope to the middle and tie the other end to the load. Lift up on the other end of the board.

Third-Class Lever

Project 16

Bottled Force

Kinetic and potential energy

Purpose Let's see if we can find a way to store "work" energy.

Overview Energy can be placed in one of two groups, kinetic energy and potential energy. Kinetic energy is the energy of work being done. It is the energy of movement. When a bowling ball is rolling down an alley, the energy of its motion is kinetic energy. Potential energy is "stored-up" energy. It means something has the ability to do work, but the energy is not being

used at the moment. If a rock is sitting high on a hilltop, it has potential energy. Because of gravity, the resting rock has the potential, or ability, to release energy. If the rock is given a push, its potential energy is easily turned into kinetic energy...as it rolls down the hillside.

We use the word mass to describe how much "stuff" an object has in it. The more mass an object has, the heavier it will be. A ping pong ball and a golf ball are both about the same size and shape, but the golf ball has more mass. Think about how much more mass a bowling ball has compared to a same size ball of cotton candy!

Hypothesis The more mass an object has, the more potential energy it will have when it is raised up, and the more kinetic energy it will have when gravity causes it to move down. This can be proven by comparing the work done by objects that are the same size and shape, but have different masses.

Procedure Lay a book about an inch (2–3 cm) thick face down on the floor. Place one end of a board about 1 foot wide by 4 feet long (30 by 120 cm) on the book, making a ramp with a gentle slope.

On the floor, about 1 foot (30 cm) from the end of the ramp, stack three or four children's wooden building blocks on top of each other.

Hold an empty 2-liter soda bottle at the top of the ramp. The bottle has potential energy (stored energy), because gravity can pull it down. Let go of the bottle, but be careful not to give it a push. Just let gravity start it rolling. Does it push over the wooden blocks? If it doesn't, remake the stack but this time use one less block. If it does, add another block to the stack. Repeat rolling the bottle and adding or taking away blocks until you find out exactly how many blocks the empty bottle will push. Do not stack the blocks more than six high. If more blocks are needed, start another pile of blocks behind the first, building a thicker wall.

Now, fill the plastic bottle with water. Be sure the cap is on tight. We are keeping the slope of the ramp CONSTANT, and the VARIABLE will be the mass of the bottle. As before, place the bottle at the top of the ramp and find out the maximum number of blocks it can knock down. Remember not to stack the blocks more than six high. Make more stacks of blocks behind the others, making the wall thicker.

When the bottle is filled with water it has more mass. Did the filled bottle have more potential and kinetic energy than when it was empty and had less mass?

Results & Conclusion   Write down the results of your experiment. Come to a conclusion as to whether or not your hypothesis was correct.

Something more

1. What can you fill the bottle with to give it more mass than water, giving it the ability to do even more work (pushing even more blocks)?

2. Change the slope (the incline) of the ramp. How does the angle of the ramp affect the bottle's kinetic and potential energy?

3. At what angle does the force into the blocks remain unchanged? At some point, the bottle will begin putting energy into the floor and no more into the blocks.

Project 17

Water Maker

Solar radiation

Purpose How could we change snow to water if the temperature was below freezing?

Overview If you were stranded in the woods or in a place where there was lots of snow and ice and you needed water to drink,. how could you get it to melt?

On a sunny day when there is snow on the ground, you may notice the snow in places starting to melt, even though the air temperature is below freezing—32 degrees Fahrenheit or 0 degrees Centigrade. Where do you see this happening? Is the ice melting because the sun is hitting it? Would the melting stop if a black plastic bag was placed over it to block the sunlight getting to it, or do you think the black color would collect even more sunlight and turn it into heat that would melt the ice faster?

> **You need**
> • 3 clear cereal bowls
> • 12 ice cubes
> • clear plastic food wrap
> • black plastic trash bag
> • sunny day when the outdoor temperature are cold
> • clock or watch
> • paper and pencil

Hypothesis Hypothesize that a black covering will absorb sunlight and the heat will make the ice melt even faster.

Procedure Place three cereal bowls outside on a sunny, but cold, day. Put four ice cubes in each bowl. Cover one bowl with clear plastic food wrap. Cover another bowl with a black plastic trash bag. Leave the third bowl uncovered. Our VARIABLE is the color of the plastic covering. Temperature, location, amount of sunlight, and the ice are held CONSTANT. We are *assuming* that either the thickness of the food wrap and the plastic bag is the same or, if not the same, that the thickness will not affect the results. What do you think will happen to each bowl of ice?

After setting the bowls outside in a sunny place, check them every fifteen minutes. Write down the time and your observations (recording what you see happening) in each bowl. If you don't see any change, write down that that there is no change.

After a time, water may appear in some of the bowls. If ice appears in more than one bowl, you can find out in which bowl the ice melted faster by "quantifying" (measur-

ing) the amount of water in each bowl or by comparing the water in each bowl after you remove the ice cubes.

Results & Conclusion   Write down the results of your experiment. Come to a conclusion as to whether or not your hypothesis was correct.

Something more

1. You may wish to continue your experiment by making covers of different colors, red, green, yellow, and others. Is there a difference in the amount of ice melted between a bowl of ice that is air tight (covered by plastic wrap) and a bowl that is not air tight (covered by a piece of colored construction paper), even if both colors are the same?
2. Does the thickness of the cover act as an insulator? Try comparing various cover thicknesses.

Project 18

An Uphill Battle

Kinetic energy and the transfer of energy

Purpose Demonstrate that energy can be transferred from object to object, and defy gravity.

Overview An energy force can travel like a wave, which means that it can be passed from one object to another. The force of the energy transferred can even be stronger than the force of gravity, so that the energy can be made to travel uphill, where it is possible for it to do more work.

You need
- 2 paper towel tubes
- string
- adhesive tape
- marbles
- 5 or 6 books
- modeling clay

Hypothesis Hypothesize that energy does pass through objects and this force can be transferred with enough strength to travel uphill.

Procedure By using adhesive tape at the sides, position a piece of string over the opening of one end of an empty paper towel tube. Fill the tube with marbles. Lay a thick book (such as a dictionary or encyclopedia volume) down. Tilt the open end of the paper towel tube filled with marbles up on the book. The other end of the tube, with the piece of string, should be at the bottom to keep the marbles from rolling out. Placing modeling clay on top of the book will hold the paper towel tube in place.

Stack four or more books face down on the table. Using modeling clay, tilt the other empty paper towel tube up onto the books, making a ramp. The bottom ends of both paper towel rolls must face each other, as shown in the drawings on the next page.

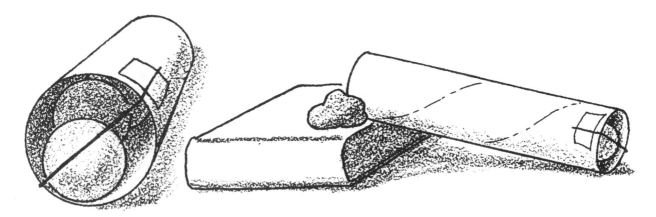

Roll a large marble (or two smaller ones together) down the steep paper towel tube. This rolling force is called kinetic energy, the energy of work being done. When the marble comes out of the bottom of the roll, it will hit the first marble in the next tube filled with marbles. The energy will pass from the rolling marble to the first marble, and then up through all of the marbles. All of the marbles in the filled tube are touching each other, and while these marbles do not move, the energy passes through them. When that energy gets to the last marble at the top, the force will move the marble. Can you see it move?

Now, repeat the experiment, this time changing the slope (the angle) of the empty striking tube. Remove two of the books to lower the slope of the tube. Keep everything else CONSTANT. The only VARIABLE is the change in the slope of the striking tube. Roll the large striking marble down the tube again. Is the weaker force of the striking marble still able to move the marble at the end of the filled tube?

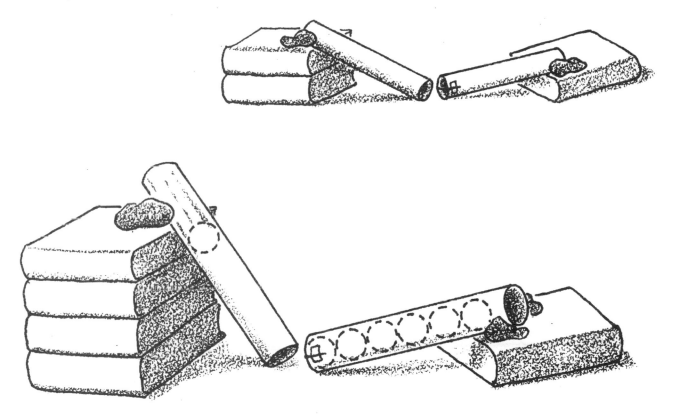

Results & Conclusion   Write down the results of your experiment. Come to a conclusion as to whether or not your hypothesis was correct.

Something more   Can you make the force so strong that it will knock the marble off the end? You can increase the force by using a larger striking marble (increases mass and momentum), and you can make the angle steeper for the striking marble. If you can get the last marble to roll out of the tube, how far can you get it to travel?

Project 19

Cellular Can

Transmitting sound by vibrating materials

Purpose To improve the sound of the traditional homemade "tin-can" toy phone.

Overview Sound is formed by an object moving back and forth, or vibrating. These vibrations move molecules in the air by first compressing them and then causing them to spread apart. In order for us to hear something vibrating, the object must be quivering with enough force for our ears to detect it (loudness). It must also be vibrating between about 16 and 20 thousand times per second, which is the frequency response of the human ear.

You can often feel the vibrations that are making a sound. Lightly touch a string on a guitar, or a harp, after it has been plucked. Lay a piece of paper on top of the speaker in a car and turn up the volume on the car radio.

A popular toy kids often make is a "telephone," put together using two metal cans and a string pulled tight between them. A small hole is punched in the bottom of each empty can through which a piece of string is tied and knotted.

You need
- an adult
- 2 small metal (soup) cans
- 2 large metal (juice) cans
- string
- hammer
- nail
- a friend

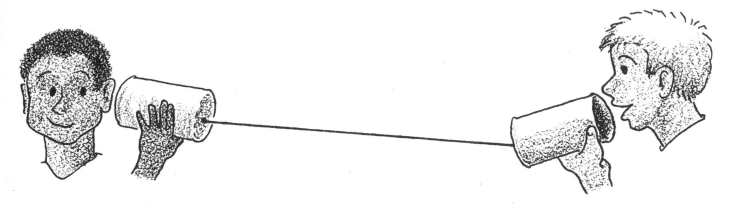

When the string is pulled tight, speaking into one can causes the can bottom to vibrate. These vibrations then travel along the taut string and vibrate the bottom of the other can, converting the vibrations back into sound. The vibrations of your voice are therefore "transmitted" to the person on the other end of the toy-telephone system, and he or she can hear you.

Can you improve on this toy, making the sound either clearer or louder?

Hypothesis   Using larger cans instead of the usual smaller "soup" cans will improve the sound transmission of a homemade string-can telephone system. (We are defining "improve" to mean either louder or clearer sound.)

Procedure   Make a set of toy "can telephones" using a piece of string and two small, empty cans with one lid removed; use cans that are the size soup usually comes in. Have an adult check the can rims to be sure there are no sharp edges that might hurt you. Also, ask the adult to punch a small hole in the middle of the bottom of each can, using a hammer and nail. Push one end of the string into the can through each hole far enough to be able to tie a knot in the string inside the can. Knot it several times, making a knot big enough so it won't pull out through the hole. Pull the string tight and hold a "secret" conversation with your friend on the other end.

Now make a set of can telephones using two larger size cans, the kind that fruit juice might come in. Is there any difference in sound quality or volume between the telephones using the larger cans?

Results & Conclusion   Write down the results of your experiment. Come to a conclusion as to whether or not your hypothesis was correct.

Something more
1. Can you think of any other kinds of materials to use instead of string that might work better? Replace the string with monofilament line (fishing line). How about thick monofilament line compared to thin line (different "pound test" strengths).
2. Try using cans that have short sides, such as ones that pineapple or tuna fish come packed in to improve sound.

Project 20

Singin' in the Shower

Acoustics: the behavior of confined sound

Purpose   To find out if different rooms in your house have different acoustics (sound qualities).

Overview   A branch of physics that studies the behavior of sound is called "acoustics." You often hear the word acoustics when someone is talking about the characteristics of sound in a particular place. Smooth, hard surfaces reflect sound waves, bouncing the waves off the object. Hardwood floors, walls, and glass are examples of things that reflect sound. Other materials absorb (soak up) sound waves instead of reflecting them, or do not reflect *all* of the sound that strikes them. These include carpeting, curtains, and couches.

When sound reflects off objects, it can create either an echo or a reverberation. An echo is a distinct repeat of a sound. When sound bounces off an object far away, an echo is often heard, such as shouting into a high cliff. The farther away the reflecting object is, the longer will be the delay between a shout and the echo. A short echo may be heard if you stand far back from the side of a brick building, such as your school might have, face the large brick or concrete wall, and give a sharp yell.

> **You need**
> - portable battery-operated cassette tape or CD player
> - portable battery-operated cassette tape recorder with built-in microphone
> - blank cassette tape
> - tape or CD with your favorite song on it
> - bathroom that is not carpeted, or an uncarpeted kitchen or long empty hall
> - a room with carpeting, window drapes, and upholstered furniture or cloth material (such as a couch or bed)
> - tape measure

The term reverberation, or simply reverb, is used to describe the sound of thousands of echoes mixed together, each with a different delay. Reverb is a wash of sounds rather than separate distinguishable sounds. It is caused by sound bouncing many times off of many different objects. You will hear a reverb effect if you talk in an empty room that has bare walls and floor. Some sound will bounce only once before reaching your ear, while other sound waves may bounce from wall to wall two, three, or several times before they reach your ear.

Reverberation and echoes can make listening difficult when a person is speaking, as a lecturer in a large hall or a pastor in a church would do; but a little reverb can make some kinds of music sound more interesting, giving them a fuller sound.

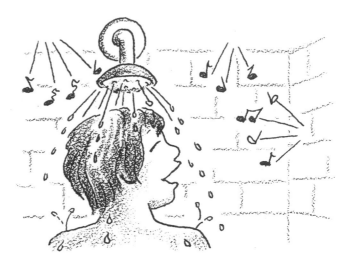

Rooms in libraries and hospitals are places where architects and builders try to reduce the reverberation of sound so as to keep the rooms quieter. Theaters are designed to keep sound from bouncing around, so that a person speaking on stage can be more easily heard and understood.

Where you live, the room that probably reflects sound the most is the bathroom; and in that room, the shower is the most reverberant. Have you noticed that effect?

Hypothesis   A sound recording made in a bathroom, which has a lot of reflective surfaces, will sound different from a recording made in a room that is heavily furnished and carpeted.

Procedure   Place a battery-operated tape recorder on the floor of a bathroom, a kitchen, or any room that does not have carpet and upholstered furniture. Place a battery-operated tape or CD player on the floor of the same room at a distance of six feet or more. Put a blank tape in the recorder. Put a music tape or CD containing your favorite song in the player. Start recording on the tape recorder, and play the song on the other player. Let the recorder and player run for about one minute. Stop them both. If you are using music tape, rewind it back to the beginning.

Set up the two machines in another room, one that is carpeted, has drapes on the windows, and has a bed with covers or has upholstered furniture, as you would find in a living room. Place the tape machines at the same distance from each other. The song, the tape machines, and the distance apart are kept CONSTANT. The VARIABLE is the environment the recorders are in. Again, start recording on the one machine and play the song on the other. After one minute, stop the tapes and rewind both of them.

Listen to the recording you made. Compare the "acoustics" or sound qualities of each recording. Write down a description of both rooms and the things that are contained in them.

Results & Conclusion   Write down the results of your experiment. Come to a conclusion as to whether or not your hypothesis was correct.

Something more
1. Lift the lid of a washing machine and give a yell, then kneel in front of a couch and yell; describe the difference in the two sounds.
2. Compare a recording made inside your house to one made outside.

Project 21

Ear of the Beholder

Pleasant sounds versus "noise"

Purpose   The purpose is to determine if there is a difference in the opinions of young people and those of an older age as to what normal-living sounds they consider pleasant and unpleasant.

Overview   Some sounds are thought of as pleasing, while others are "noise." The difference between a pleasant and an unpleasant sound may be in the mind of the listener. Hitting a fence with a stick may be noise, but if you walk along a picket fence and run the stick against it, the regular repetitive sound may be pleasing to you but more than annoying to the homeowner who is sitting on his porch listening to you scratching his fence!

> **You need**
> • paper
> • pencil
> • clipboard
> • a day of listening
> • 10 teenagers
> • 10 older adults
> • use of a computer and printer with a word processing program (optional) or access to a copy machine

Sometimes pleasant sounds can be unpleasant, depending on the circumstances. The sound of a telephone ringing during the day can be pleasant; it could be a friend's expected call. The sound of a telephone ringing at 3 o'clock in the morning can be annoying and maybe alarming. The sound of a doorbell ringing during the day may be pleasant, but in the middle of the night may be a worry or even frightening.

Hypothesis   When young people and older people are surveyed as to a list of sounds they think are pleasant or unpleasant, the results will be different between the different age groups.

Procedure   For one whole day, pay attention to all of the sounds you hear. Carry paper, a pencil, and a clipboard to make a list of all the daily sounds around your home and neighborhood. Some sounds you may not have paid much attention to before, for example: toast popping up in a toaster, a door chime, a church bell, popcorn popping, a car horn, the crackling of a fire in a fireplace, the telephone ringing, birds chirping, the next door neighbor's dog barking, an umpire or referee blowing a whistle during a sporting event, insects buzzing in your ear, the screech of car brakes, and a friend blowing air across the top of a soda bottle.

After you have made up your list, think about other sounds you sometimes hear; a babbling brook, rap music playing loudly on a boom box, someone sitting at a table "drumming" a pencil in beat with a song in his head.

Take your notes and make a list of fifty sounds. Alongside the items, make three columns and head them PLEASANT, UNPLEASANT, and NO RESPONSE. Make twenty copies of your list (or use a computer to make up and run off copies).

Give your survey to ten adults and ten teenagers (write their names at the top). Ask each one to mark one column for each item on the list. Tally their marked answers, totalling the PLEASANT, UNPLEASANT and NO RESPONSE items for all of the adults, then do the same for the teens. Which age group found which sounds to be pleasant or unpleasant? Are there sounds that they agreed on?

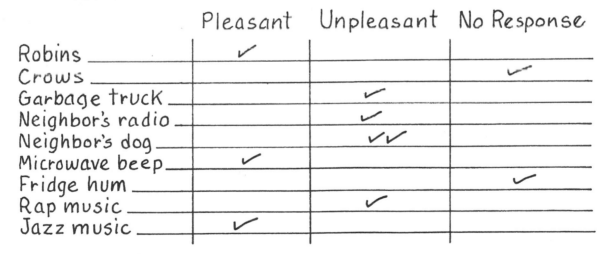

Mr. Brown

	Pleasant	Unpleasant	No Response
Robins	✔		
Crows			✔
Garbage truck		✔	
Neighbor's radio		✔	
Neighbor's dog		✔✔	
Microwave beep	✔		
Fridge hum			✔
Rap music		✔	
Jazz music	✔		

Results & Conclusion Write down the results of your experiment. Come to a conclusion as to whether or not your hypothesis was correct.

Something more Rearrange the results of your survey based on male and female, rather than on age group.

Project 22

The Sound of Time

Amplifying sound

Purpose Sound is sometimes faint and hard to hear, how can it be amplified to improve hearing?

Overview Sound waves traveling through the air can be gathered to make them louder. They can also be directed (focused) in one direction to make them louder. Have you ever seen a band shell behind a large orchestra playing outside?

One way that sound is directed is by using a megaphone. A megaphone is a horn-shaped device used to increase the sound of a person's voice. Cheerleaders at a football game or the lifeguard at a beach often use megaphones. Early record players, made before the invention of electronic amplifiers, used such horns to make the music louder for listeners.

You need
- 2 standard-size sheets of construction paper
- adhesive tape
- a clock that "ticks" or watch with an alarm
- modeling clay
- outdoor picnic table or chair
- a friend

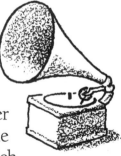

A megaphone works in reverse, too. It can gather sound and allow a person to hear weaker sounds better. Think about the shape of your outer ear, which is responsible for gathering sound. At a football game on television, you might notice a technician standing on the sidelines holding a large curved dish? This parabolic dish has a microphone attached to pick up what the players are saying.

Hypothesis   Hypothesize that by using a paper megaphone at the source and one at your ear, you will be able to hear a sound louder than one that, without these devices, you are able to hear not at all or just barely.

Procedure   Roll a piece of paper into the shape of a horn and use adhesive tape to keep it in place. Make another horn so you have two of them.

On a quiet, calm day, go outside and find a place (table, bench, stoop) to mount your sound source. You will need a clock that has an audible "tick" or a watch with an alarm, and some modeling clay in order to stand the watch on its side.

Make sure the clock is ticking or turn the watch alarm on so that it produces its beeping sound. Move away as far as you can from the clock or watch until you can just barely hear the sound. Then stop and hold one of the horns to your ear. Can you hear the sound any better?

Now have a friend hold the other horn in front of the clock or the watch with the large end toward you. Does the beeping get even louder? The source of the sound is CONSTANT; our VARIABLES are two megaphone devices.

Note: If you do this experiment on a windy day, it could affect your results. When doing science projects, it's important to control *all* the variables, which means keep all things constant (the same) except for those that are changed on purpose. If the wind is constant, that is, if the wind speed and direction are the same when you listen with and without the horn, maybe the results of the experiment can be trusted. But if the wind is gusting or swirling, it will very likely change the results.

Results & Conclusion   Write down the results of your experiment. Come to a conclusion as to whether or not your hypothesis was correct.

Something more   Does frequency (cycles per second) have an effect on the ability of the megaphones to amplify a sound? Use a music-instrument keyboard and compare a low note to a high note, both with and without the aid of the megaphones.

Project 23

Blown Away

Fluidics: air flow around shapes

Purpose Determining how air flows around objects could sometimes be very helpful to know.

Overview It is important to understand how moving air behaves. Airplanes lift off the ground because of the way air travels past the wings, which have a special shape. When two tall buildings are close together, wind can speed up as it travels between them, causing a windy condition that may be undesirable.

Hypothesis The shape of an object affects how moving air flows around it.

> **You need**
> • hand-held hair dryer
> • coffee mug
> • nail, about 2 inches (5cm)
> • hammer
> • small block of wood
> • scissors
> • piece of yarn
> • a pint or quart milk carton
> • paper and pencil

Procedure Prove that the shape of the object affects how moving air flows around it.

Using a hammer, drive a 2-inch-long (5 cm) nail partially into a small piece of wood as shown. Near the head of the nail, tie a piece of yarn tightly onto the nail. Cut the yarn so it is about 3 inches (7 cm) long. This will be our "air-flow indicator."

Place a round coffee mug on a table. Place the air-flow indicator about 2 inches behind it.

Hold a hand-held hair dryer in front of the mug and turn it on at the highest speed. Use a cool setting if it has one. The fast moving air splits, hugs the mug as it travels around it, and comes together behind the mug. The yarn will stand out straight like a flag or a windsock in a strong breeze, showing that air is moving quickly.

Move the air-flow indicator to various

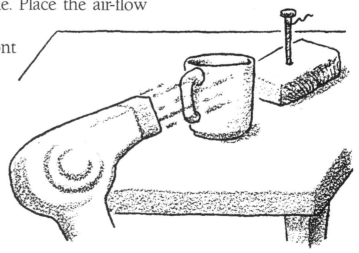

spots along the side of and behind the mug to find places where the air is moving. On a piece of paper, draw a diagram of the mug and hair dryer. Make it a view looking down from the top of the mug. Mark spots on the paper to show where there is moving air, as detected by your air-flow indicator. Use arrows to show the direction of its flow. Do you see a pattern of the air flowing around the mug?

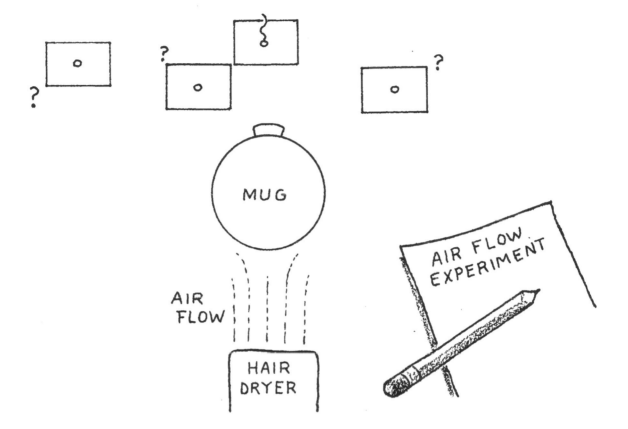

The velocity of the moving air will be held CONSTANT. Changing the shape of the object in the stream of air flow will be the VARIABLE. What if you turn the handle of the mug to one side or the other?

(Note: Since fast moving air flows around rounded objects and meets behind it, maybe hiding behind a tree or telephone pole to block wind is not as effective as you might think!)

Now, replace the mug with a rectangular-shaped object, such as a pint or quart carton of milk. Again, move the yarn air-flow indicator around the carton and draw a diagram showing the air flow around it.

Results & Conclusion   Write down the results of your experiment. Come to a conclusion as to whether or not your hypothesis was correct.

Something more   Making the piece of yarn longer will allow it to become an indictor of air speed. The farther out the yarn is blown, the stronger the air flow. Now you can use the device to compare the strength of air flow in addition to direction.

Project 24

Balancing Act

Objects at rest tend to stay at rest

Purpose   We want to find out if objects at rest will remain at rest if balanced forces are applied.

Overview   Sir Isaac Newton, an English mathematician who lived from 1642 to 1727, discovered some important principles of natural science. One of the laws of nature he recognized is that "objects at rest tend to stay at rest," which means that things that are not moving will stay that way unless an "unbalanced force" pushes (or pulls) on them. An unbalanced force is a push on an object that is stronger in one direction than a push from the opposite direction.

Balanced forces occur when equal forces coming from opposite directions are applied to an object. A book lying on a table has balanced forces acting upon it; the table is pushing up and gravity is pulling down. If the forces are equal, but they are not at exactly opposite angles, the resulting force will be unbalanced. In other words, the book may slide down the slanted table.

Hypothesis   Objects at rest tend to stay at rest when balanced forces are applied to them.

Procedure   To put together a device to prove, or disprove, the above hypothesis, ask an adult to help by cutting a wooden dowel into three 3-inch-long (7 cm) cylinders. The dowel should be at least 1 to 1½ inches in diameter. If the dowel is purchased at a hardware store or hobby shop, the sales clerk may offer to cut the dowel to size for you.

With a hammer, tap a small nail into the top of each dowel at exactly the middle. To find the middle accurately, you can use two pieces of string or thread. Lay one piece across the end of the dowel and another straight across it, at a 90-degree angle. The point where the two pieces cross each other is the middle.

Tie one end of a 3-foot-long (90 cm) piece of string onto the nail of one of the dowels. Do the same for the other two dowels.

> **You need**
> - an adult
> - a 9-inch (21 cm) wooden dowel, an inch or more (3 cm) in diameter
> - hand wood saw
> - 3 small nails
> - hammer
> - thin string
> - scissors
> - broom handle or yardstick
> - 2 chairs of equal height

Place two chairs equal in height back to back (such as matching kitchen or dining room table chairs), but separate them by about 3 feet. Lay a broom handle across the top of the chair backs (a yardstick or any long, stiff pole will also work).

Tie the loose end of one of the dowels to the center of the broom handle, so that the dowel hangs down, but does not touch the floor.

Hang another dowel on the right and another one on the left of the first dowel, so they are side by side. Tie the ends to the broom handle so that all three dowels are hanging straight and just touching alongside each other when they are not moving.

Take the left dowel in your left hand and the right dowel in your right hand. Pull them both away from the center hanging dowel until they are each about a foot (30 cm) away from the center dowel. Let the dowels in each hand go at exactly the same moment (this may take some practice), so they will both hit the center dowel together.

If only one dowel should swing into the center dowel (which is at rest), the "unbalanced force" will push the center dowel and make it swing, too. But, if both swinging dowels apply *equal* force in *opposite* directions, the experiment should result in a balanced force on the center dowel so that it remains at rest.

Results & Conclusion   Write down the results of your experiment. Come to a conclusion as to whether or not your hypothesis was correct.

Something more   Show balanced and unbalanced forces in the game of "tug of war." Tie a ribbon onto the center of a long rope. Place a brick or some object on the ground to mark a spot, and lay the rope over it, with the ribbon on top of the brick. Have several friends grab one end of the rope and several grab the other for a game of tug of war. When equal pulling force is on both sides, the ribbon will stay hovered over the brick. When the pulling force becomes unbalanced, the ribbon will move toward the friends who are exerting the stronger total force.

Project 25

Floating Along

Buoyancy: the ability to stay up

Purpose Let's figure out how come things that are heavier than water can float.

Overview Why do things float? An object may float in water if it is light and weighs less than water. A ping pong ball will float because it weighs less than water.

But why does a heavy boat float? Big ships are made out of steel, and steel is much heavier than water.

The "buoyancy" of an object is its ability to float on the surface of water (or any fluid). Water gives an upward push on any object in it. The amount of force pushing upward is equal to the

You need
- modeling clay
- large bowl
- water
- small kitchen food scale (gram weight scale)
- small bowl or cup
- kitchen measuring cup that has a pour spout
- thin piece of thread

weight of the water that the object "displaces" (takes the place of). So, if a boat or ship is designed to displace an amount of water that weighs more than the boat, it will be able to float. For something to be buoyant, its shape is very important.

Hypothesis An object that is heavier than water can be made to float.

Procedure Fill a large bowl with water, but don't fill it all the way to the top. Take a small amount of modeling clay and use your hands to roll it into a ball that is about two inches (5 cm) in diameter. Place the ball on the surface of the water and let go. The clay ball is heavier than water. Does it float or sink?

Take the ball out of the water. Use your hands to mold the same clay into the shape of a small boat. It should have a flat bottom and sides. Now, place the boat on the surface of the water. It floats, even though it is the same amount of clay. While holding the weight (mass) of the clay CONSTANT, the VARIABLE has been its change in shape.

You can take this project further by capturing and weighing the water that is displaced by the ball of clay. To do this you will need a kitchen measuring cup that has a pour spout. Set a small bowl or cup under the spout to catch the water that spills out.

Fill the measuring cup with water until water begins to spill out of the spout. When the water stops overflowing, empty the small bowl that caught the water. Dry it out.

Shape the clay into a ball. Tie a piece of thin thread onto the ball and slowly lower it into the water. The water it displaces will spill out into the bowl.

When the water stops overflowing, remove the bowl and weigh it on a small kitchen food scale. Write down the weight.

Dry the bowl and weigh it. This is to get the "tare weight," the weight of the container that had been holding the water. Subtract this tare weight from the weight of the bowl with the water in it. The difference is the weight of the displaced water.

Also weigh the clay ball. Compare the weight of the clay ball to the weight of the water that it displaced.

Results & Conclusion   Write down the results of your experiment. Come to a conclusion as to whether or not your hypothesis was correct.

Something more   Weigh the amount of water the clay boat displaces. Compare it to the amount of water the same clay in the shape of a ball displaces. Do you think the weight of the water displaced by the boat will be less than the weight of the water displaced by the ball?

Project 26

Flying in Circles

Air friction: sometimes good, sometimes not

Purpose Set up some demonstrations showing how air friction (resistance) affects falling objects.

Overview When we see or hear the word "friction" we usually think about it taking place between two surfaces. As Project 4 explained, friction is the resistance to motion when two things rub together. Friction can also occur, however, between air and any object that moves through it. Air friction causes resistance, which pushes against the object.

You need
- plastic trash bag
- thread or thin string
- 2 metal washers
- 2 standard size sheets of typing paper
- scissors
- pencil

Car manufacturers try to design cars that are "streamlined," which means the cars are shaped to let air flow smoothly around them without offering much resistance. This reduction in air friction is desirable because it then takes less energy to move the car. This saves on fuel. The car will get more "miles (or kilometers) per gallon"—go farther on less gasoline. Streamlined cars save gasoline (an important resource), the cost of gasoline at the fuel pump, and the car will also put less pollution into the air.

Designers of aircraft, just like the designers of cars, try to make their crafts have as little resistance to the air as possible, but sometimes, air friction is desirable. When sky divers jump out of an airplane, their parachutes need to have a lot of air resistance to slow their descent and land them safely on the ground.

Hypothesis Air friction can be increased and decreased simply by changing an object's shape.

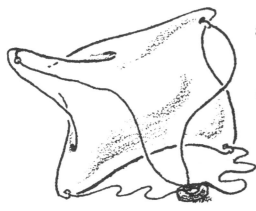

Procedure First, we'll compare how air friction affects two sheets of typing paper, keeping the material and height CONSTANT and shape the only VARIABLE. Fold one of them into a paper airplane shape and crumple the other into a ball. Stand in a clear area in a room. Stretch your arms out to your sides, holding the airplane nose down in one hand and the paper ball at the height of the airplane nose in the other. At exactly the same time, drop them.

Next, with a scissors, cut a section of plastic the size of the sheet of paper out of a plastic bag. Take a pencil or nail and carefully poke a small hole near the edge at each corner. Cut four pieces of thread or thin string, each 2 feet in length. Tie one end of each piece to each hole in the plastic. Tie the loose ends of the four threads to a metal washer. The washer is our sky diver, and the plastic bag is his parachute.

Now, cut another piece of plastic the same size, put a washer inside, and wrap and tie it into a bundle using 8 feet of the same thread or string. Holding the parachute and wrapped plastic out so they are the same height from the floor, let them go. Here again, shape is the only VARIABLE.

In order to really demonstrate the difference streamlining makes, do one more drop. In one hand, hold your sky diver. In the other hand, hold the paper airplane, with its nose facing the ground. Outstretch and raise your arms so the bottom of the sky diver is at the same distance from the ground as the nose of the airplane. Let go of both objects at the same time. Does the sky diver's parachute encounter much more resistance, from friction with the air, than the streamlined paper airplane?

Results & Conclusion Write down the results of your experiments. Come to a conclusion as to whether or not your hypothesis was correct.

Something more Would it make a difference if the shape of the parachute was square or round instead of rectangular? Would another shape have even more air resistance?

Project 27

Pop a Treat

The physics of popcorn

Purpose   If it's moisture that causes popcorn to pop, is it measurable?

Overview   Smelling the distinct aroma of freshly popped popcorn probably brings to mind a trip to a movie theater to see a great movie. Popcorn is a healthy and tasty food.

The kernels of corn used to make popcorn have a strong airtight outer covering that seals moisture inside. When heat is applied, the moisture turns into superheated steam. This pressure build-up eventually bursts through the outer coat, expanding the contents to about thirty times its original size. Of course, the moisture escapes, but how much? Is it possible to quantify the amount lost during the popping process?

Hypothesis   Hypothesize that the moisture in a serving size of popcorn kernels, which causes the kernels to pop when heated and escapes as steam, is not measurable—even though we are able to see it as steam.

Procedure   Tape a small paper plate onto each end of a ruler, to keep them from moving, and make a balance scale by placing a pencil at the middle of the ruler with the paper plates.

Put fifty kernels of unpopped popcorn in a small brown paper lunch bag. Put fifty kernels in another lunch bag. Place one bag at the center of each paper plate.

Try to balance the two bags by adding kernels to the lighter of the bags. When you have the two bags balanced, remove one of the bags, being careful not to bump the ruler, pencil, or the other bag.

Fold the very top of the bag closed, to keep the popping kernels inside the bag. Place the bag in a microwave oven and have an adult help you set the controls and pop the kernels in the bag.

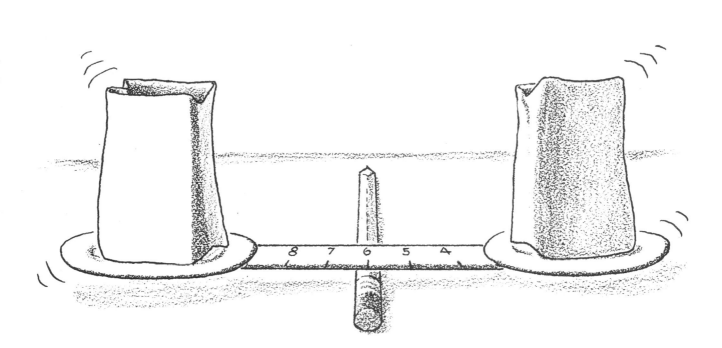

When the popping is done, take the bag out (use oven mitts to prevent being burned) and open the bag. Do you see steam coming out? This is visual proof that there was water in the popcorn kernels, which was heated to steam and has now just been released. The question is whether or not the lost moisture is measurable.

Place the bag back onto the balance scale, positioning the bag in the center of the plate as before. The number of kernels in the microwaved bag has remained CONSTANT, and the moisture content is the VARIABLE.

It's okay if there are a few kernels that remain unpopped. Can you detect any noticeable difference in weight? If so, was the bag of popcorn lighter than the bag of unpopped kernels? Eat the popcorn. Enjoy!

Results & Conclusion Write down the results of your experiment. Come to a conclusion as to whether or not your hypothesis was correct.

Something more The reason some kernels of corn don't pop can be because there is a scratch or break in their airtight coating, which lets steam escape slowly and prevents sufficient pressure build-up. You may want to try an experiment and see if you can prevent kernels from popping by first scratching their surface or carefully poking them with the sharp point of a drawing compass (use caution when handling sharp tools).

Project 28

Unwelcome Gusts

Comparing and measuring wind strengths

Purpose Over a week's time, we want to determine which day had the strongest gust of wind.

Overview People have harnessed the powerful force of wind to help them do many useful things. Windmills have been used to pump water and make electricity. Boats can sail around the world by using sails whose large surface area captures the wind's power.

But, sometimes this wind force works against us. On extremely windy and gusty days, huge bridges are sometimes closed to vehicles with large surface areas, such as tractor trailers and motor homes, because of the danger. Playing beachball or volleyball in a strong wind can either help your team or hinder it, depending on whether or not you are downwind.

On a windy day, have you ever tried to carry a big piece of plywood or poster board, or a small bag holding only something light, like a greeting card? Have you ever helped your parents put garbage cans out at the curb on a windy trash day? If no one is home when the trash is collected, you could come home to find the cans scattered all over. The empty cans, being lighter without trash, are easily blown all over the yard and even into the street, causing a hazard to motorists, by strong gusts of wind. Can you think of other times when the force of the wind is not welcome, or is even harmful?

Wind velocity is important in air travel. In physics, the word "velocity" means both speed and direction. Airports use wind socks to get a relative indication of wind direction and speed.

You can build a simple weather instrument to detect wind gusts and get a comparative indication of their strength by using water-filled paper cups.

Hypothesis Hypothesize that you can put together a simple device that will allow you to determine the force of the strongest wind gust of the day.

> **You need**
> - 7 medium size (8 oz.) plastic or foam cups
> - piece of board
> - 2 cinder blocks/trash cans
> - a wide open area, away from buildings
> - water
> - kitchen measuring cup
> - heavy mug or old pot

Procedure Find an open area away from buildings or other structures that might block the wind. A spot in your own backyard would be good, if place is available.

Set two cinder blocks upright on the ground several feet apart. If you don't have cinder blocks, you can use same-size milk crates, or buckets or trash cans turned upside down. Across the top of any such "risers," lay a long piece of wood.

Set seven 8-ounce paper cups (or plastic cups) in a row on the board. Leave one empty and, using a measuring cup, pour 1 ounce of water in the second cup, 2 ounces in the next, 3 in the next, continuing up to 7 ounces. If your measuring cup is marked in milliliters, use increments of 50 (that is, 50 ml, 100ml, 150 ml, 200 ml, and so on).

On the board or in an open place nearby, place a heavy mug or old pot. This will be used to capture any rainfall for the day. If you find rain in it, don't record that day's results because they won't be valid.

At the end of each day, observe which cups have blown off the board. Write down how many cups blew off. The lighter cups are more sensitive to the force of the wind.

The next day, set them up again, and refill the cups with water (some water will probably have evaporated). The contents and positions of all of the cups must be kept CONSTANT. The wind gusts should be the only VARIABLE in the project. At the end of the day, record which cups have blown off the board.

Do this every day for a week or two, or as long as you wish. Look at your recorded observations for each day. Did your paper-cup system work as a weather instrument, to allow comparisons of the strongest gusts of wind that occurred each day?

Results & Conclusion Write down the results of your experiment. Come to a conclusion as to whether or not your hypothesis was correct.

Something more What if you found a heavy and a light cup knocked over, but one in the middle still standing? Do you think the results of that day should not be used, as there may have been interference from squirrels, birds, or other animals seeking water?

Project 29

A Lopsided Pinwheel

Balancing points of oddly shaped objects

Purpose How to discover the balancing point of irregularly shaped objects.

Overview The balancing point of a square or circle is easy to find. You just need to measure. It's right at the center of it, as long as the object is basically two-dimensional or flat; that is, doesn't have a significant amount of depth (a rock or glob of clay is definitely third-dimensional).

The balancing point, however, is not always in the center. It's the point at which, if an axle is placed through it, an object can be spun like a wheel and it will stop at a different spot every time. Is it possible to find the balancing point of something flat, but that has an odd unsymmetrical shape?

You need
- a push-pin
- piece of thin cardboard, oak tag, or standard size construction paper
- cardboard box, standing at least a foot high
- scissors
- thread
- metal washer
- pencil

Hypothesis Hypothesize that you can find the balancing point of an irregularly shaped piece of stiff paper or cardboard.

Procedure Draw an irregular shape on a piece of thin cardboard, oak tag, or construction paper and cut it out with scissors. You may draw something like the six-sided shape shown here. Although the cut-out shape will be really three-dimensional (it has *some* depth, or thickness, as well as length and width), for the purposes of our project, this third dimension is so small we can assume it will not affect the results.

Set a cardboard box on a table. Near one edge at the top of your shape, place a push-pin through your object and into the side of the cardboard box. If you don't have a cardboard box, you can use a cork bulletin board or a piece of plywood, as long as it is kept perpendicular to the ground

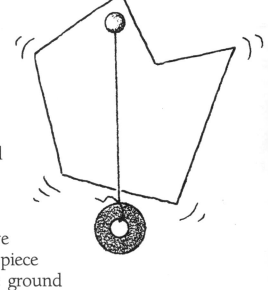

(standing up straight). Do not push the pin in all the way. Swing the object back and forth to be sure it can move freely.

Tie one end of a piece of thread onto the push-pin, and tie a metal washer to the other end. If you don't have a washer, use any object as a weight to cause the string to hang straight down (a large paper clip, for example). We are going to use gravity to make sure the line we draw will be perpendicular (at a 90-degree angle) to the ground. The thread must also hang straight down and be hanging freely, not hung up on anything. With a pencil, draw a line tracing the path of the string across the object.

Next, take out the push-pin, turn the object about 90 degrees (it does not have to be turned an exact amount), and push the push-pin into a point near the top edge and through to the cardboard box. Again, hang a thread from the push-pin and draw a line tracing the path of the string as it falls across the surface of the object. The shape of the object is our VARIABLE; gravity and the line with the washer hanging down are CONSTANT.

Where the two lines intersect (where they cross) is the balancing point of the shape. How do you know this?

First, you can verify it by turning the object once again and repeating the hanging of the thread from the push-pin, drawing a third line. That line, too, should cross at the same spot where the other two meet. Other similar lines will also cross there.

Second, prove that this is the balancing point by pushing a push-pin through that intersecting point and into a stick, making a pinwheel-like toy. Give the object a spin a number of times. Each time it should stop at a different point.

Results & Conclusion Write down the results of your experiment. Come to a conclusion as to whether or not your hypothesis was correct.

Something more
1. Place numbers all around the edges of your shape to make a spinning "fortune wheel" game (in this case "Hexagon of Fortune"). You and your friends can guess what number it will stop on when given a good spin. Try turning your object into a pinwheel by bending up some of the edges to catch the wind.
2. Draw and cut out more objects, each with a different numbers of sides or some irregular shapes. Find the balancing point of each object and prove it. Or you and your friends can guess where the balancing points will be and find out, using the methods above, who comes closest.

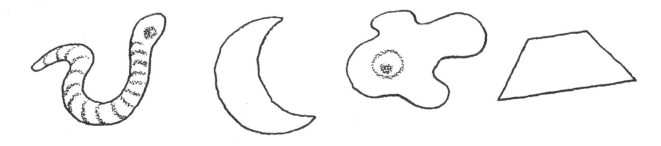

Project 30

Get the Point

Changes affecting surface areas

Purpose Studying surface area differences.

Overview A famous trick from long-ago India was for a person to lay down on a bed of nails. Ouch! How could they do that without the nails sticking right into them? Physics, that's how!

The term "surface area" refers to the total area of an object exposed to the space surrounding it. Imagine a sailboat trying to get anywhere with a sail only 2 feet (60 cm) by 2 feet. Its surface area would be only 4 square feet (360 cm square), certainly not enough to catch any meaningful wind. The large surface area of a big sail is needed to really capture and harness wind power.

> **You need**
> - an adult with a saw (to cut the wood)
> - 4 squares of soft pine board, 11 inches (28 cm) by about ½ inch (1.25 cm) thick
> - 104 roofing nails, 1½ to 2 inches (4–5 cm) long
> - hammer
> - ruler
> - pencil

Let's get back to the man on the bed of nails. The surface area of the pointed end of a large nail is certainly not very big. But, what if there were a thousand nails all fairly close together? If a person lies gently on the bed, the weight of the person could be distributed over the total surface area of all the nail points. Lying on a bed of nails doesn't sound like something you would want to do, but with the help of physics and "surface area," it seems not as painful as you might think.

Women wearing very tiny high (spike) heels, perhaps only ¼ inch or 1 centimeter square, have been known to break the toes of a dance partner. Imagine even an average-sized grown woman concentrating almost all her weight on two such little square heels! Each of those little heels would have a very powerful downward pressure!

We can expose the "bed of nails" trick as a fraud by proving that when weight is distributed over a lot of nails compared to just a few, there is less force per nail tip because of the total increase in surface area of the combined tips.

Hypothesis When a weight is divided over a greater surface area (many nails), each nail will make a smaller impression in a piece of wood under it.

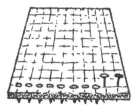

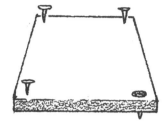

Procedure Have an adult cut some soft pine board into 11-inch (28 cm) squares. The boards should be thin, no bigger than about ½-inch thick.

With a ruler, make marks all the way around the edge of one of the pieces of wood at 1-inch (2.5 cm) increments. Then draw lines from the points on one side to the other point on the opposite side, making a grid of 1-inch (2.5 cm) squares across the surface of the board.

Lay the board outside on the ground. Using a hammer, pound a flat head roofing nail at each spot where lines intersect. There will be 100 intersecting points. The nails must be long enough to go through the board and stick out the other side, extending through at least a half inch (about 1.25 cm). When pounding the nails in, they will stick harmlessly into the ground.

Lay another piece of pine wood on a hard surface, such as a sidewalk or paved driveway. On top of it, lay the board with nails, with the nails facing down. Next, place another pine board on top. This board will help insure that none of the nails will back out when weight is put on them. Now stand on the boards. Jump up and down once. Remove the top two boards and observe the nail depressions made in the bottom board.

Now we want to keep the weight CONSTANT, but use less nails, making the surface area the VARIABLE. So, lay the fourth board on the ground and pound four nails into it, one near each corner, perhaps one inch (2 cm) in from the sides.

Turn the board with the nail depressions in it over so its smooth side is facing up. Lay the board with only four nails on top of it, with the nails facing down. As before, lay the other piece of pine on top to keep the nails from backing out. Then stand on the boards and jump up and down once.

Remove the boards. Examine the board with the nail depressions in it. Are the nail depression marks deeper when all of your weight was distributed by only four nails compared to when the surface area of 100 nails distributed your weight?

Results & Conclusion Write down the results of your experiment. Come to a conclusion as to whether or not your hypothesis was correct.

Something more Try measuring the depth of the impressions. Use a toothpick.

Project 31

Break the Beam

Exploring some characteristics of light

Purpose The purpose is to understand how a light security system works.

Overview Can you see a beam of light? You can certainly see a lit light bulb, the sun, a candle's flame, or any source of light. You can also see objects because light is shining on them. But, you are normally not able to see the light beam itself. The path of the beam through the air can, however, be seen by filling the air with tiny particles so the light will reflect off them.

In a dark room, lay a flashlight on a table and shake several facial tissues in the air. The light will reflect (bounce) off of the tiny particles of tissue, allowing you to see the path of the light beam. Similarly, have you ever seen the rays of sunlight shine from behind breaks in clouds? Have you ever seen the light beams coming from the headlights you were riding in on a dark, foggy evening?

Because we can't actually see a beam of light, some home security systems use light to detect if someone walks through a room. A light source shines into an electronic device that detects the light. Everything is fine as long as light is shining on it. But when the light beam is interrupted when someone walks between the light source and the security device, it senses that the light has gone out, and it sounds an alarm. To make the light less detectable by anyone, a red filter is used to reduce the light reflected by any tiny particles that may be in the air.

You need
- lamp
- hand-held mirror
- modeling clay
- small piece of cardboard
- flashlight
- several facial tissues
- dark room
- a friend
- table

Hypothesis We can detect a person walking in another room using a light source and a mirror.

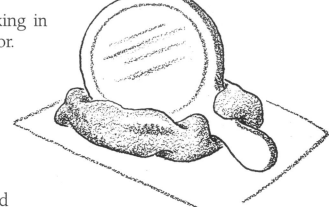

Procedure Set up a demonstration of how a home security system might work. Lay a hand mirror on its side on a piece of cardboard. With modeling clay, build a base around it so the mirror will stand up by itself, as shown.

Set the mirror on a table or bookshelf, and adjust it so that the light from a lamp is reflected into another room. It should be a room that you can make fairly dark. Stand against the wall and look out the door at the mirror. Have a friend adjust the mirror until you can see the lamp in it. Then you know the mirror is lined up. This demonstrates another characteristic about light; light travels in a straight line unless something interferes.

Go into the dark room and close the door until it is only open enough to let a slit of light in to shine against the wall opposite the door.

Watch the light on the wall as you have your friend walk around the room. Can you detect when your friend steps in the path of the light?

Results & Conclusion Write down the results of your experiment. Come to a conclusion as to whether or not your hypothesis was correct.

Something more You can tell the direction of a person walking through the room by adding a second mirror a few feet to either side of the first mirror, so that a second light spot shines on the wall. Then, if a person walks by, they will break one beam before the other. The beam that is broken first tells you in which direction the person is walking.

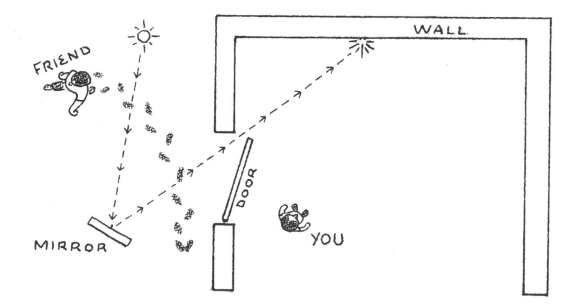

Project 32
Now Soft / Now LOUD
Amplifying sound through reflection

Purpose To understand and demonstrate how sound vibrations can be amplified through the use of a "sounding board."

Overview A sound something makes can often be made louder by focusing it, reflecting it, or making a second object vibrate or resonate.

Did you ever see a cheerleader using a megaphone? A megaphone is a large horn with a small opening at one end and a large opening at the other. Sound is focused by a megaphone. Speaking into the end with the small opening makes the user's voice louder. It can also be used in reverse, holding the small opening end up to your ear to collect distant or faint sounds and make them louder (see Project 22).

Another way to amplify sound (make it louder) is to cause the vibrations to be expanded to another object. The hollow wooden body of an acoustic guitar gives the vibrating strings of the instrument more volume and better tone quality.

A third way is to use something called a "sounding board." The sounding board in a piano reflects the sounds of the vibrating piano strings, making the sound louder.

Hypothesis Hypothesize that sounds from a vibrating body can be amplified by adding a sounding board.

Procedure Take a bobby pin, the flat metal hairpin used to hold ones hair in place, and bend one of the prongs out until it makes a 90-degree angle (so that the bobby pin forms an "L" shape). Normally, one prong of a bobby pin is straight and the other contains zigzag curves. With

You need
• an adult
• bobby pin
• small block of wood
• staple gun with staples, or hammer and small nails

your thumb and index finger, tightly hold the bobby pin near the bottom of the straight prong, holding only about ½ inch (1 cm) of the prong. Use the thumb of your other hand to sharply flick the end of the curved prong, making a downward stroke. Listen closely to hear the faint sound.

We are going to keep the vibrating bobby pin CONSTANT; the VARIABLE will be the addition of a block of wood to act as the sounding board.

Now, ask an adult to help you mount the bobby pin on a small block of wood. The wood should be about 1 inch (2.5 cm) thick by 5 or 6 inches (13–15 cm) square or rectangular. (The exact length and width are not important.) Mount the bobby pin by placing two staples from a staple gun or two small nails (brads) to attach the bobby pin to the wood block. The nails can be tapped partway into the wood then bent over the pin, using a hammer. About ½ inch (1 cm) of the straight prong should be mounted flush against the wood block, and the rest of the bobby pin should rise above the block. If staples are used, give them a tap with the hammer to ensure that the bobby pin is firmly touching the wood. Sharply flick the end of the curved prong with your thumb, with a downward stroke. Is the tone you heard before louder now?

Results & Conclusion   Write down the results of your experiment. Come to a conclusion as to whether or not your hypothesis was correct.

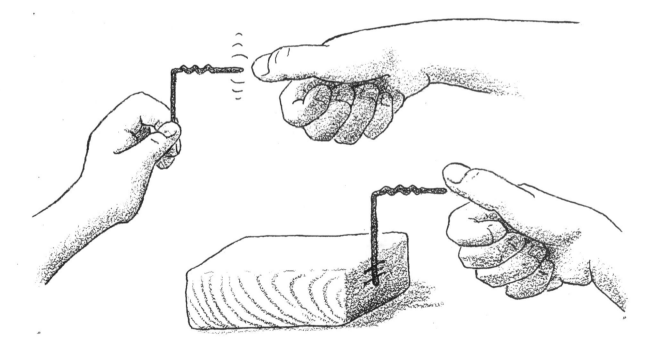

Something more
1. Will increasing the size of the wood block amplify the sound even more? At what size does further increase make no difference in the sound?
2. Can you make the sound even louder by resting the board on a large wooden table, holding it tightly against a wall, or against a large metal pan?

Project 33

Plant First Aid

Collecting light to benefit plants

Purpose To increase the amount of light available to a houseplant not getting direct sunlight.

Overview We have a houseplant that sits on a table where it does not get much light. The table is not near any windows in the room, and the room is sometimes a little dark because of an overhang on the porch outside the only window in the room.

Plants need light. To help the plant out, we constructed a simple device to capture light and bathe the plant in more light than it would normally receive. Since light travels in straight lines but can be reflected, we constructed a three-sided cardboard reflector to collect the light that missed the plant and bounce it back toward the plant.

You need
- 2 large pieces of cardboard
- 2 similar house plants, or batches of growing seedlings
- aluminum foil
- adhesive tape
- black paint or construction paper
- art supplies (optional)
- scissors
- 2 months time

Hypothesis A plant exposed to more light, by having a reflector surrounding it, will grow better than one that does not have the reflector. (We consider "growing better" to mean the plant will have either bigger or more leaves, or will be healthier and grow taller.)

Procedure Obtain two houseplants of the same type that are similar in appearance: about the same height and number and size of leaves. (You could also grow your own.)

Find a location in your home that gets good light but not direct sunlight.

Take a large piece of cardboard, big enough to go around three sides of one houseplant and as tall as the plant—or cut one from a larger piece of cardboard. Bend the cardboard in two places to make "wings" that can be angled so the cardboard can stand up by itself, as shown. In the same way, construct a second three-sided cardboard stand.

To make the stands more attractive, since the project will take some time, use your art supplies to draw, paint, or otherwise decorate the backs of the shields, the convex side that will be facing away from the plants.

Line the inside (the concave side) of one of the cardboard stands with aluminum foil (shiny side out); use pieces of adhesive tape or glue to keep it in place. Paint the inside of the other cardboard stand black, or cover it with black construction paper.

Set up the plants and shields in the location you've found for the project. Position the wings of the cardboard stands mostly open, but closed enough so the cardboard will stand on its own. Place one plant on the concave inside of each cardboard stand so that the two plants will receive the same amount of normal room lighting and not be shaded.

The plants must also be cared for equally. Be sure to water the plants regularly. When you water them, each plant must receive the identical amount of water.

The CONSTANTS in this project are the room temperature and the amount of water they receive. The VARIABLE is the amount of reflected light each plant receives.

Results & Conclusion Write down the results of your experiment. Come to a conclusion as to whether or not your hypothesis was correct.

Something more If your reflector worked, can it be made smaller so that it is more attractive in a room, yet still gives the same results? Try cutting the size of the reflector in half.

Project 34

Hot Light

Comparing waste heat from bulbs

Purpose Let's see if using a higher wattage incandescent light bulb to increase brightness also produces more waste heat energy.

Overview If we want more light in a room, we could change the wattage of the light bulbs used in lamps in the room. Although brightness is really measured in units called "lumens," light bulbs are often sold by wattage, the rating is listed on the package. A "watt" is a unit of electrical energy. Bulbs that are brighter require more electrical energy to use them, so they have a higher wattage rating, based on consumption. Consumers have a good idea as to the brightness of a bulb by comparing its wattage rating. A 15-watt bulb is commonly used in refrigerators. A 25-watt bulb might be found in a small bedside lamp. A 75- or 100-watt bulb is used in overhead fixtures in a kitchen or workroom or in large living room lamps where strong light for reading is needed. Other wattage bulbs available are 40 and 60 watts.

> **You need**
> • an adult
> • lamp
> • thermometer
> • wooden stick
> • clear adhesive tape
> • 25-watt incandescent light bulb
> • 100-watt incandescent light bulb
> • ruler
> • clock or watch
> • pencil and paper

While we get more light from higher wattage light bulbs, we may also be getting more of something else that is unwanted—heat. Some energy is given off from light bulbs in the form of heat, which is wasted. Does a 100-watt incandescent light bulb give off more waste heat than a 25-watt light bulb?

Hypothesis An incandescent bulb that produces more brightness (has a higher wattage rating) also produces more heat.

Procedure Safety is always the first concern when doing any science project. Because light bulbs can get very hot and it is important to be very careful working with electricity, have an adult unplug and plug in the lamp and remove the light bulbs as needed.

Using clear adhesive tape, attach a thermometer to a wooden stick or dowel, and position the sensitive tip of the thermometer 2 inches (5 cm) from the end of the stick.

You can use any kind of stick: an ice-pop stick, a tongue depressor (available at your local pharmacy), a wooden dowel (found in a hardware store or hobby shop), or a small twig from a tree. The adhesive tape should be clear or positioned so that it does not interfere with your reading of the numbers on the thermometer.

Find a lamp on which the lampshade can be easily removed. Once an adult unplugs the lamp from the wall outlet, unscrew the light bulb and replace it with a 25-watt bulb. Carefully re-insert the lamp's plug and turn the lamp on.

Hold the wood and thermometer device against the side of the bulb, as shown. Only the wood, which does not conduct heat, should be touching the bulb. Do not touch the thermometer glass itself while you are holding the device, and do not hold the device over the bulb, to avoid excess heat building up in your hand.

Holding the thermometer device parallel to the tabletop, wait for three minutes. Then, read the temperature and write it down.

Turn the lamp off. Wait about ten minutes for the bulb to cool off and for the thermometer to return to room temperature. Remember, a glass light bulb may be hot, but not look hot.

Our CONSTANT is the distance the thermometer is from the bulb surface, and the VARIABLE is the wattage of the bulbs.

Carefully unplug the lamp from the electric receptacle. Unscrew the 25-watt bulb and replace it with a 100-watt bulb. Plug the lamp back in; turn it on and again hold the thermometer by the bulb and wait three minutes. Be very careful around the bulb. It is hot! Write down the temperature.

Compare the temperature readings from the two bulbs. Does the bulb that is brighter also produce more heat?

Results & Conclusion   Write down the results of your experiment. Come to a conclusion as to whether or not your hypothesis was correct.

Something more

1. Find the temperature of bulbs having 25, 50, 75, and 100 watts and see if there is a mathematical relationship between the wattage and the temperature.

2. Find a lamp that uses a fluorescent bulb, in your home or at a friend's, and use your thermometer device to see if it gives off as much heat as an incandescent bulb of the same wattage rating.

71

Project 35

Crash!

The relationship between mass and force

Purpose It often happens that objects that are at rest, that is, not moving, are hit by moving objects and forced to move. What happens when the objects struck have different masses?

Overview Sir Isaac Newton did experiments to find the mathematical relationship between the mass of an object and how fast it moves when a given force strikes it. Mass is how much "stuff" an object is made up of. Newton found that the larger the mass of an object, the smaller will be its movement when a given force is applied.

Imagine a soccer ball filled with air and another one that is filled with sand. If you kicked each soccer ball with the same amount of force, the ball with more mass (filled with sand) would not move as far as the one with less mass (filled with air). You might hurt your foot on the ball with more mass, too!

You need
- an adult
- ladder
- soccer ball
- bowling ball
- golf ball
- several thick hardback books
- 2 wooden boards, 2-by-4-inch by 8 feet (240 cm) long
- hammer
- 6 long nails
- ruler

Hypothesis A soccer ball will move farther than a bowling ball when the same force is applied to each.

Procedure Let's strike a bowling ball (in place of a sand-filled soccer ball) and a soccer ball with the same force and measure how far they move. If Sir Isaac Newton is right, the soccer ball, which has less mass than the bowling ball, will move farther. (Be very careful handling the bowling ball. It could hurt your foot if it should fall on it. Have an adult help you if the ball is too heavy for you to handle safely.)

We need to have a force that will be exactly the same every time, so we can be sure each ball is struck with an identical force. This is our CONSTANT. To do this, construct a ramp with the long two 2-by-4 inch wooden boards, making a "V" shape. Rolling a golf ball down the "V" channel will cause it to strike whatever object is at the bottom

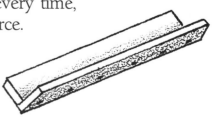

of the ramp with the same force every time. If we let go of the golf ball at the same place on the ramp each time, the force of gravity will ensure that the ball is rolling at the same speed every time it reaches the bottom of the ramp.

Nail two 2-by-4 8-foot (240 cm) long pieces of lumber together, making a "V" shaped channel. This will be our wooden ramp.

Outside, set up a ladder. Rest one end of the ramp on the third or fourth rung of the ladder. At the ground end of the ramp, place a bowling ball so that it is touching the end of the ramp. The ground must be flat and level.

You'll get the most action if the bowling ball is struck in its middle, some small distance above the ground. Place books underneath the end of the ramp to raise it until it is positioned at the middle of the bowling ball. You may need to place a few books along the sides of the ramp to keep the "V" shape facing up.

Pick a spot along the ramp to let go of a golf ball and start it rolling down the ramp. To get the most speed out of the ball, you can let it go from the high end of the ramp. Be sure, however, that you let the ball go from the same spot every time. Also, don't give the golf ball a "push" start, because you would not give it an even push every time. Just let go of the ball and gravity will start it rolling.

If the bowling ball moves when it is hit, use a ruler to measure how far it moved.

Now we want to see how far a soccer ball, which has much less mass, will roll when the same force is applied to it. The different masses of the two balls is the VARIABLE in our project.

Release the golf ball. If the soccer ball moves a lot, it may be easier to use a tape measure, yardstick, or meterstick than a ruler to measure the distance it rolled.

If neither ball moved, increase the slope of the ramp by moving up one rung on the ladder, giving more speed, hence force, to the rolling ball.

Results & Conclusion Write down the results of your experiment. Come to a conclusion as to whether or not your hypothesis was correct.

Something more Repeat this experiment using different balls at the bottom of the ramp; try a baseball, basketball, and a tennis ball. Can you predict which one will roll farthest?

Project 36
Bigger Water
Temperature's expansion/contraction effects

Purpose   What happens when water freezes?

Overview   Many things expand or contract when they change temperature. Have you ever noticed, when standing by a railroad track, why there are gaps in the rails at certain intervals? The spaces in the rails allow them room to "grow," in case they expand; otherwise, the rails would buckle. Track engineers know exactly how big the gaps should be to allow for this rail expansion.

> **You need**
> • empty metal soup can
> • water
> • use of a freezer
> • small bowl

If you have electric baseboard heat in your home, you may have heard the crackling sounds it makes when the metal fins heat up or cool down. That's because the fins are expanding and contracting. Gaps in the roadway of bridges are also there to allow for expansion and contraction from temperature changes.

Hypothesis   The same amount of water takes up more space when it is frozen.

Procedure   Fill an empty soup can with water. (Be careful of sharp can edges.) Set the can in a small bowl, and place it in the freezer section of a refrigerator. The bowl will catch any water that might spill from the can. Add more water if necessary so the water level in the can is at the very top. Leave the can of water in the freezer overnight.

In this experiment, the quantity of water is being held CONSTANT, and the temperature is our VARIABLE.

Take the can of ice out of the freezer in the morning. The volume of water that fit into the can when it was a liquid is now too big for the can. The ice has risen above the top of the can because of the expanding water and its push against the bottom of the can.

Results & Conclusion   Write down the results of your experiment. Come to a conclusion as to whether or not your hypothesis was correct.

Something more   Quantify how much more volume the ice takes up than it did as a liquid by using displacement, which is explained in Project 25. Hold the can tightly to melt the ice slightly around the edges of the can so the ice will come out as one block. Dip the block in a container of water and measure the water that it displaces.

Project 37

Look at the Sound

Light speed faster than sound speed

Purpose Prove that sound travels much slower than light through the medium of air.

You need
- an adult with a car
- a straight, mile-long, open stretch of road

Overview Lightning streaks across the sky. Wait! It's only then you hear the rumble of thunder. You hear a jet plane flying high overhead. You look up immediately, but it's already way past you. You're at a baseball stadium, far from home plate. The batter's ready. There's the pitch! It's a hit, and the ball starts to soar. Then, you hear the crack of the ball against the bat!

Why do we see things before we hear them? Could it be because speed of light is faster than speed of sound through the air? (Actually, light travels 186,000 miles per second, so when something happens we see it almost instantly. Sound travels through the air much more slowly, only about 1,100 feet per second—764 miles per hour at 32 degrees Fahrenheit—slightly faster at higher temperatures.)

Hypothesis We can prove that sound travels more slowly than light through air.

Procedure Find an open stretch of road at least a mile long. Some evening, as you stand safely off to the side of the road, have an adult with a car drive one mile away, then turn the car around and turn the headlights on and "beep" the horn at the exact same time. Do you see the flash of the headlights before you hear the sound of the horn?

The CONSTANTS are the distance and the time the headlights are turned on and the horn is sounded. The VARIABLE is the time it takes for the light and sound to go from car to you.

Results & Conclusion Write down the results of your experiment. Come to a conclusion as to whether or not your hypothesis was correct.

Something more
1. Does wind direction have an effect on the travel of sound through the air?
2. Does sound take so long to reach your ears that, in the above experiment, the driver could blink the lights on and off several times before you even hear the horn?
3. What things can transmit sound but not light, or light but not sound? Think about a supernova, a railroad track.

Project 38

Work = Force × Distance

The wedge, a simple machine

Purpose Show the relationship between the distance a wedge is moved forward and the height an object sitting on top of the wedge is raised.

Overview A "wedge" is one of those "simple machines" we talked about. A wedge is an object in the shape of a triangle. A doorstop and the metal head of an axe are examples of wedges.

When an axe or chopping maul is used to split firewood, the worker swings the tool over a large distance to strike the wood with great force. That force is turned into the small distance covered by the wedge, as the axe moves down into the wood to split it. In science, "work" is a measurement equal to "force" times "distance."

When a force is applied to a wedge, the force moves the wedge forward, but it also moves anything resting on top of the wedge into an upward direction (at a 90-degree angle to the forward movement of the wedge).

A wedge can be used to lift very heavy objects a short distance. House movers sometimes use wedges between the sill plate and the foundation to raise a house up so steel girders can be slid under it.

Hypothesis Using a wedge increases the amount of force in a perpendicular direction, but we pay for it in a decrease in distance.

Procedure Draw a right triangle on a thick piece of cardboard. Make the triangle about 2 inches (5 cm) tall by about 12 inches (30 cm) in length. The hypotenuse of the triangle will form a long, gently sloping ramp. Use scissors to cut the triangle out.

> **You need**
> - a wide strip of thick cardboard, about 12 inches (30 cm) long
> - two rulers
> - small, light cardboard box (shoebox or a similar size)
> - scissors
> - pencil
> - paper
> - 1 or more heavy books

Place a small, light box on a table. A shoebox would be perfect. At one end of the box, stack one or two heavy books. That will keep the box from sliding.

At the other end, place your cutout wedge so that its pointed tip just slips under the box. Lay a ruler alongside the box, with the zero mark on the ruler at the edge of the box where the wedge touches the box. The length the ruler should face away from the wedge (running parallel to the side of the box).

Push on the wedge so that it slides 2 inches (5 cm) under the box. Use another ruler to measure how high the end of the box is raised above the table.

On a piece of paper, draw two vertical columns. Label the heading on one column DISTANCE WEDGE MOVED and the other column HEIGHT RAISED. Write the measurement under the first heading and the distance the box was raised in the second column.

The CONSTANT in this project is the incline (the slope) of the wedge, the box it is lifting, and the force applied. The VARIABLE is the distance the wedge is moved inward and the height it pushes up on the box.

Now push the box forward another inch or centimeter, and record the height raised. Continue to push the wedge under the box at each increment, until the top of the wedge is reached. Write down the distance and the height for each move increment.

Results & Conclusion   Write down the results of your experiment. Come to a conclusion as to whether or not your hypothesis was correct.

Something more   A wedge doorstop is a stationary wedge that is applying a force equal to the force needed to keep the door from closing.

A nail is also a wedge. Can you imagine pushing something into a piece of wood that doesn't come to a point? The smaller the nail, the easier it is to wedge into the wood, because it has less wood material to push out of the way. Try pushing a nail into a piece of wood by hand. Then try pushing a thumbtack with a head on it into the same piece of wood. Is the thumbtack much easier to push in?

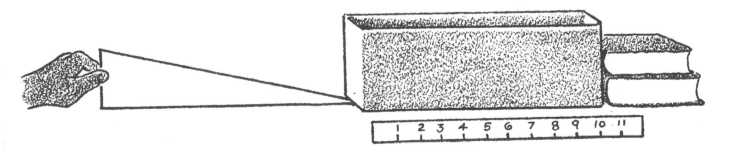

77

Project 39

Water Shooter

Compressing a fluid

Purpose   Let's learn a little about "hydraulics."

Overview   Hydraulics, from the Greek word meaning "about water," is the study of liquid in motion. A part of hydraulics deals with compressing a liquid, used in machines where great pushing or lifting strength is needed. A force pushing on any part of an enclosed liquid creates an equal pressure per unit of area on everything the liquid touches. By using a system of pistons (cylindrical containers filled with a liquid), great force can be achieved.

You need
- squeeze bottle with spout top
- a piece of heavy board
- water
- ruler

Hypothesis   When water is compressed and forced to flow out of an opening, the velocity of the water will be much greater if the opening is small than if it is large.

Procedure   Outside, remove the spout and fill the squeeze bottle with water. Set the board on its side and tilt the container against it. Let water leak out until it stops; the angle will keep most of the water inside. Then with your hand, strike the side of the bottle, forcing water out through the opening. Watch how far the stream of water shoots.

   Again, fill the squeeze bottle with water, set it against the board and let the water leak out. This time, screw on the spout. Keeping the volume of water CONSTANT (by tilting) as well as the striking force used on the bottle, our VARIABLE will be the diameter of the opening through which the water escapes. The spout makes the opening much smaller. Strike the side of the container again, using the same amount of force as before. Does the stream of water travel farther with the spout on? Does that mean the velocity of the water coming out was greater?

Results & Conclusion   Write down the results of your experiment. Come to a conclusion as to whether or not your hypothesis was correct.

Something more   Compare the amount of water coming out of the bottle with the spout off and with it on. Use a measuring cup to quantify the volume of water in the bottle before and after each strike. Perhaps, with the spout on, less water is coming out? Did you strike the bottle with the same force each time? Can you find a way to be sure?

Project 40
Siphon Fun
Water drains to own level

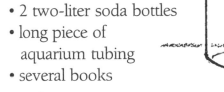

You need
- 2 two-liter soda bottles
- long piece of aquarium tubing
- several books
- water

Purpose   Discover how a siphon works.

Overview   A hose or tube can be used to create a "siphon," a device that drains liquids. A siphon will drain liquid from a higher to a lower level, even if it first has to travel uphill! A hose placed with one end underwater in an above-ground swimming pool and the rest draped over the side and down to the ground will drain water out of the pool. To get the flow started, it may be necessary to suck on the low end of the hose. Once the flow begins, gravity pulls the liquid down, creating a vacuum in the tube that draws the liquid up and through the tube.

Hypothesis   Water can be made to rise above its level through the use of a siphon.

Procedure   Stack books on a table. Fill a clear plastic bottle with water and place it on the books. Set an empty bottle near it, but on a lower level—*not* on the books.

Insert one end of plastic tubing, available at pet shops, into the bottle of water. Be sure there is space around the tubing in the bottle's neck (if it is air-tight, the siphon will not work). The end of the tubing should touch the bottom of the bottle. Suck on the other end of the tubing, as you would a soda straw, to get the water flowing, then insert the tube into the empty bottle. Push the tubing in so it touches the bottom of the bottle.

Held CONSTANT is gravity, the volume of water, the lack of air in the tube system, and the bottles. The motion of the water from higher to lower bottle is the VARIABLE.

Results & Conclusion   Write down the results of your experiment. Come to a conclusion as to whether or not your hypothesis was correct.

Something more
1. Once the lower bottle is filled, do you think that the siphon would work in reverse? What if the full bottle was now raised higher than the one on the books?
2. Does it take the same amount of time for the bottle to empty each time the experiment is done, so that it could be used as a sort of "water clock"?

Project 41
Whirring Button
Torque: changing the direction of force

Purpose   Learning about torque and changing the direction of a force. A force pulling outward can be changed into a force at a 90-degree angle and made to do work, causing a button on a string to spin.

Overview   A force can push or pull an object along a straight line. When a force is used to rotate, or turn, an object, physicists have a special name for the force. They call it "torque."

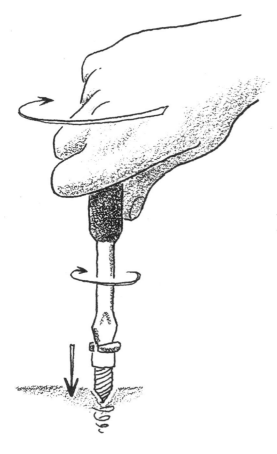

Torque has a direction associated with it. Look at these drawings. As your wrist turns a screwdriver *clockwise*, the force rotates the screw *forward* into the wood. That torque, or force, is in a different direction from the turning force applied by your wrist. The same is true when a wrench is used to turn a nut on a bolt.

Hypothesis   A change in the length of string used to cause a button to spin will cause a change in the rate of the rotating button.

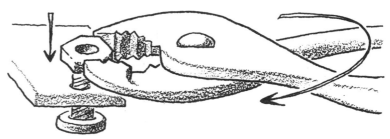

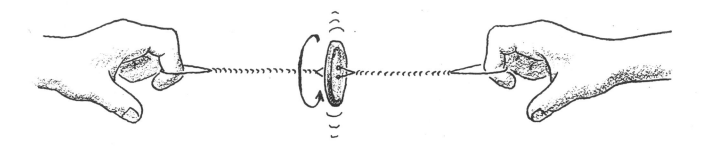

Procedure Loop a piece of cotton thread through a hole in a large button, and then back through another hole. Cut the thread to a length of 2 feet and then tie the ends together.

In the same way, loop a piece of cotton thread through an identical button, but cut the length of this piece of thread to 4 feet, and then tie the ends together.

Hold the looped thread between two hands, moving the button to the middle of the thread. Move your hands together so that the button hangs down, and use a circular motion with your hands to swing the button around and around, winding up the thread, until it has wrapped many, many times. Have a friend hold the second button and longer string, and wind the thread in a similar way, giving it the same number of turns as you do. Note: If your flat buttons have four holes, instead of two, you can place the thread through two diagonal holes, as shown.

At the same time, both you and your friend pull outward with each hand, and the thread will begin to unwrap, causing the buttons to spin. The outward pulling force of your hands is being changed into the rotating motion of the button.

The mass of the buttons and the number of winding turns was held CONSTANT, and the length of the string was the VARIABLE. We are "assuming" that the force pulling outward on both button devices is equal. Repeat the test, exchanging whirring buttons.

Does one device spin faster than the other? Do the spinning buttons set up vibrations in the air causing a tone that is audible—you can hear? If so, is the pitch higher on one than the other? Is the higher pitch coming from the button that is rotating faster?

Results & Conclusion Write down the results of your experiment. Come to a conclusion as to whether or not your hypothesis was correct.

Something more

1. Experiment using different threads: does string or monofilament line (fishing line) work better or worse in rotating a button? What effect does changing the size of the button have on the speed of rotation?

2. Can you think of other places in our daily lives where torque force is used?

Project 42

Toy Story

Torque-energy storage in a homemade toy

Purpose Determine the optimum (best) number of rubber band windings for an ice-pop toy.

Overview This simple old-time homemade toy demonstrates three concepts of physics: potential energy, torque, and elasticity. A stick is placed inside the loop of a rubber band and turned, so the band is wound up. The wound band has "potential" energy; that is, stored energy that, when released, becomes "kinetic" energy. As with the whirring button in Project 41, the concept of "torque," a twisting force, is also demonstrated.

> **You need**
> • 3 ice-cream pop sticks
> • rubber band
> • hardbound book

The more a rubber band is twisted, the more potential energy it can store. But is there a point where additional winding does no more good, or can even be bad?

Hypothesis A certain number of elastic band turns will give consistently best results.

Procedure Hold one ice-cream stick in each hand and place a rubber band around them. In the middle of the rubber band, place a third stick.

Turn the center stick, winding the rubber band tighter and tighter. Count the number of turns. After it is wound many times, carefully place the toy under a book on a table. Lift up the book and the toy will dance.

Wind the toy again, this time making five more turns than before. The elastic band and the sticks are CONSTANTS, the number of turns is the VARIABLE. Does the toy dance better with the extra twists?

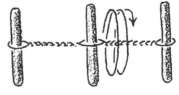

Experiment by adding more and more windings. Is there a limit to the torque energy stored in the elastic band beyond which there is no benefit?

Results & Conclusion Write down the results of your experiment. Come to a conclusion about your hypothesis.

Something more At what point does the toy have potential energy? Kinetic energy? What effect does using different rubber bands (longer or thicker) have on the energy stored? (For fun, surprise family and friends with hidden, dancing torque toys of all sizes!)

Project 43

Walls Don't Move

Newton's 3rd: action and reaction

Purpose   Understanding the physics concept of work through force and distance.

Overview   Newton's Third Law of Motion explains that if one body exerts a force on a second body (action), the second body exerts an equal force in the opposite direction (reaction), back to the first body. In other words: For every action there is an equal and opposite reaction. Imagine yourself and a friend floating in a swimming pool, in two inner tubes touching each other. If you push your friend's tube away, the force will push back and yours will move, too. Astronauts in orbiting spacecraft push against walls in order to float backward. Work, in physics, is defined as: work = force × distance. Since anything multiplied by zero is zero, if either force or distance is zero, then work equals zero.

You need
- roller or in-line skates
- flat surface and wall

Hypothesis   According to the physics definition of work, it is possible that no work is done, no matter how much effort is used.

Procedure   Wearing shoes that won't slip, stand facing a wall and push on it as hard as you can. The wall doesn't move, and you don't move. Force × distance = zero. No work was done (no distance, no work). Even if you pushed long enough to work up a sweat and you are out of breath from trying, technically no work was done because nothing moved. Although effort or force was applied, no work was accomplished!

Now, put on roller or in-line skates. Face the wall and push against it. Again, the wall doesn't move, but a force equal and opposite to yours pushes back, causing you to roll away from the wall. The wheels reduced friction, you rolled…and work was done!

Results & Conclusion   Write down the results of your experiment. Come to a conclusion as to whether or not your hypothesis was correct.

Something more   Measure the distance you traveled from the wall by pushing on it. If you push harder (apply more force), does the distance increase? Is the distance you travel in relation to the force you apply?

Project 44

Sandwater

Erosion caused by friction of glacial ice

Purpose   Without taking hundreds of years, we can demonstrate the erosive effect of a glacier.

Overview   There are places in the world where the heat of summer is not enough to completely melt the snow that has fallen during the winter. Some of the snow that does melt into water makes its way deep into the snow, where it turns into ice. As each year goes by, more and more layers of ice are built up, and the underlying layers become pressed tighter and tighter. A "glacier" is formed.

You need
• sand
• 2 paper cups
• water
• use of a freezer
• an old painted board
• a pair of heavy gloves or
  mittens

Glaciers can be as small as a few acres. Ice-sheet glaciers can spread out like continents. The Antarctic ice sheet is about five million square miles in area, about one and a quarter times the area of the United States.

Often, glaciers form on the side of mountains. When they become heavy enough, they begin to slide down the mountain slope. This movement is very slow, perhaps only a few inches or centimeters a day. But as this tremendous mass moves, it picks up stones and other materials that it then rubs against the rocks and ground underneath, causing erosion. The incredible force of glacier movement can change the way the terrain looks; it can even shear off the sides of mountains and change a "V" shaped valley into a "U" shape. The friction and force of glacier ice acts like sandpaper, eroding away even very hard rocks.

Hypothesis   Sand embedded in ice will cause abrasions in a surface it rubs against.

Procedure   Fill a paper cup with water. Place it in a freezer. Line the bottom of another paper cup with sand. Fill the cup with water, and place it in a freezer.

When the water has completely turned to ice in the two cups, take them out of the freezer. Remove the ice from the cups by turning the cups upside down until they slide out.

Pull on a pair of gloves or mittens to protect your hands. Pick up the blocks and turn them back over (sand side down). Hold a block of ice in each hand. Pushing fairly hard, rub both blocks of ice with equal strokes on an old painted board. After several minutes, examine the area of the board that the two ice blocks were rubbed against.

Does the area rubbed by the "sand-ice" block look different from the area rubbed with the smooth ice? Are there scratches? Rub your fingers gently over the board. Does one feel more abrasive than the other?

Results & Conclusion   Write down the results of your experiment. Come to a conclusion as to whether or not your hypothesis was correct.

Something more
1. As you increase the rubbing pressure, do the gouged-out, scratched areas get deeper? Do you think a heavier glacier would cause more erosion than one of less mass?
2. Make a comparison of the abrasion caused by a block of ice with fine sand particles embedded in it to one with coarse sand.

Project 45

Back in Shape

The characteristics of elasticity

Purpose   The purpose is to find the effect of weathering and stretching on the elasticity of a rubber band.

Overview   One characteristic of a material is a measure of its "elasticity." Elasticity is the ability of a material to return to its original shape after it has been stretched or pressed together. Balloons and rubber bands are very elastic. Can you think of other examples?

Have you ever seen a rubber band wrapped around a newspaper or something else that has been outside for some time? Have you ever seen an old rubber band that has been wrapped around something in an attic for many years? Can you observe cracks in the bands? Are they able to be stretched? Do they snap back to a smaller shape?

Hypothesis   After a rubber band has been stretched and exposed to outdoor weathering elements for several days, the elasticity of the rubber band will be affected, and it will not return to its original shape.

Procedure   For this project, we'll need a device to test the elasticity of material.

Gather two identical rubber bands, a large empty milk carton, and two paper clips. Lay the two rubber bands on top of each other to check that they are as equal in shape as possible. The rubber band will be our CONSTANT. Our VARIABLE will be to stretch one rubber band and expose it to weather, while the other will remain unstretched and indoors, protected from weathering.

Bend open the paper clips to make "S" shaped hooks. Carefully poke a hole in the top of a large empty milk carton, and insert a paper clip hook through it.

Hang another paper clip hook on an outdoor clothesline and drape a rubber band on the bottom part of the paper clip hook. Hang the milk carton on the rubber band using the paper clip hook in the carton, as shown on the next page.

Pour some water into the milk carton. This will create a weight to stretch the rubber band. We want to stretch the rubber band a lot, but not to the breaking point. Since rubber bands are not all the same, we can't tell you how much water to add to the carton. (Science is not always like a food recipe!) Slowly add more water to the milk carton until it looks like the rubber band is well stretched, but not in danger of breaking.

Leave the rubber band stretching device hanging on the clothes line for three or four days. Then, carefully, take the device apart and remove the rubber band. Line it up next to the other identical one that you had put aside indoors and compare them. Has the stretched one changed shape? Is it able to completely return to it's original shape?

Inspect the stretched rubber band closely. Are you able to see any cracks, discolorations, or other signs of deterioration caused by the experiment?

Results & Conclusion Write down the results of your experiment. Come to a conclusion as to whether or not your hypothesis was correct.

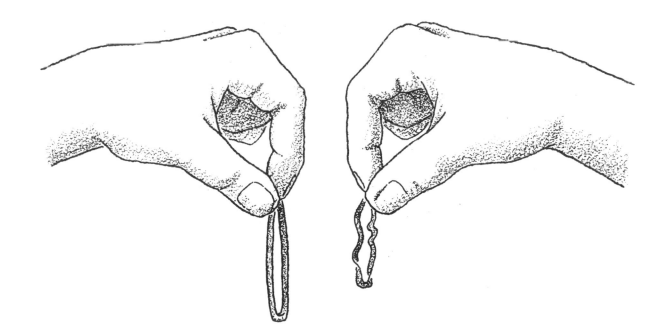

Something more Can you use your stretching device to test the elasticity of other materials or objects?

Project 46

Don't Fret

Changing pitch by varying string tension

Purpose An understanding of frequency change through an increase in tension on a stretched rubber band.

Overview Some musical instruments have strings that vibrate in order to make sounds: guitar, banjo, violin, harp, to name a few. A string is pulled taut and "plucked." The plucking force applied to the string causes it to vibrate. The mechanical vibrations are converted to sound by our ears.

As a string is pulled tighter by tuning pegs on the instrument, the frequency of its vibration increases and the notes sound higher in pitch. On a guitar and bass guitar, strings are kept under tension as they are stretched across a long wooden neck. The instrument has "frets" that the musician places his fingers against to shorten the strings that are played, changing the pitch of the sound.

Hypothesis The more a rubber band is stretched, the higher the pitch it makes when it's made to vibrate.

Procedure Hammer a nail partway into a piece of wood about a foot (30 cm) long and several inches or centimeters wide. The wood should be ½-inch (1 cm) thick or more. Don't pound the nail all the way in. Set the board near the edge of a table.

With scissors, cut a large rubber band in half. Tie one end to the nail and the other end to the handle of a small toy bucket, the kind a young child might use in a sandbox or to play with on the beach.

Drape the rubber band and bucket over the board lengthwise and off the end of the table, so that the bucket hangs in the air.

Put a few stones in the bucket to put tension on the rubber band. With your finger, pluck the length of rubber band that is stretched across the board. Like the string on a guitar or violin, the band will vibrate, producing a sound that you can hear.

> **You need**
> - large rubber band
> - piece of board about 1 foot (30 cm) long
> - nail
> - small play bucket
> - different size stones
> - scissors
> - hammer

Add a few stones to the bucket to stretch the rubber band even more. Pluck the band again.

Is the note sound higher or lower than before? The rubber band is the CONSTANT in the experiment, and the amount it is stretched is the VARIABLE.

Results & Conclusion   Write down the results of your experiment. Come to a conclusion as to whether or not your hypothesis was correct.

Something more

1. Can you add or take away stones from the bucket to match a note from the rubber band to a piano, organ, or guitar in your home? You must have a "musical ear" to be able to tell when a note on your instrument is in tune with one on the real instrument. If you do not have a good ear for matching notes, have a friend who is taking music lessons help you. Someone who has music training may do best at matching notes.

2. Challenge Project: Our musical scale consists of 12 notes in an octave. Take your project further by constructing 13 rubber band systems, each one stretched to be in tune with a different piano note (the 13th note being one octave higher than the beginning one).

Project 47

Space Saver

The concept of volume

Purpose Learn that an object's weight (mass) can remain the same, regardless of its shape.

You need
• balloons
• bucket

Overview Space in landfills, where towns and cities get rid of their trash and garbage, is limited. When things can be compressed so they take up less room, landfill space is preserved.

The volume of an object is a measure of how much space it takes up. Tin cans and aluminum soda cans, those that are not returned to stores for the deposit, are among the items collected and recycled so the metal can be reused. Cans in a recycling truck can take up a lot of room. If they were crushed, they would have far less volume, so they would take up much less space. But, they would still weigh the same.

Hypothesis Objects can be made to have less volume but still keep the same mass.

Procedure Instead of crushing soda cans to demonstrate volume, we'll use balloons. Inflate some, making each balloon about the size of a soda can. In a bucket, put in as many balloons as you can fit.

Now, take the balloons out of the bucket and lay them on a table. Count them. Lay the same number of uninflated balloons next to them. Compare the difference in the volume of the uninflated balloons to the inflated balloons—much different, right?—yet both weigh about the same. (Since air actually has some weight, the inflated balloons *do* weigh more, but the amount is negligible—not an important, easily measured, difference.)

Results & Conclusion Write down the results of your experiment. Come to a conclusion as to whether or not your hypothesis was correct.

Something more
1. Quantify the volume by how much water can be stored in a balloon. Does each balloon hold one quart? Buckets are usually sold by the number of quarts they can hold. Does a 12-quart bucket hold 12 quart-filled water balloons?
2. How many uninflated balloons can be put inside one balloon?

WORDS TO KNOW
A Glossary

acceleration   an increase in speed. To accelerate means to go faster. Physicists define acceleration as a measure of the rate of change of velocity over time.

acoustics   the study of how sound behaves, usually in rooms, halls, and auditoriums. The volume, quality, and amount of reverberation of sound are often important acoustic characteristics about a room.

assumption   When experimenting, scientists often make assumptions that certain things are true. An assumption is something that is *believed* to be true.

buoyancy   the buoyancy of an object is its ability to float on the surface of water (or any fluid). Water gives an upward push on any object in it. The amount of force pushing up is equal to the weight of the water that the object "displaces" (takes the place of).

calibrate   to make a correction or adjustment, often to a measuring device. When two thermometers are being used in an experiment, it is important that both are reporting temperatures accurately. When they are in the same place, they should both read the same temperature. If one is reading higher than the other, it must be noted and the difference in temperature must be added or subtracted from the other one in all experiments where the temperatures on the two thermometers are being compared.

centrifugal force   a force that pushes outward when an object is moving in a curve.

conductor   a material that makes an easy path. Metal is a good conductor of heat, making an easy path for heat to be carried along.

diameter   the distance across an object measured as if a straight line was drawn from one side to the other through the middle.

echo   a distinct repeat of a sound. If the ear hears two sounds that are the same but the second sound is at least 65 milliseconds later than the first sound, the brain will interpret them as an echo or as separate sounds. An echo is caused by sound reflecting off of a surface and returning back to the ear.

elasticity   the material's ability to be stretched or compressed and then return to shape.

force   a push or a pull on an object. It is an action that can change the motion of a body. Sources of a force include gravity, electricity, magnetism, and friction.

frequency   Referring to radio waves, the "frequency" of an electromagnetic wave is the number of times it "vibrates," or "cycles," per second.

friction   Friction is the resistance to motion when two things rub together. Friction keeps a car on the road. It makes your hands warm when you rub them together quickly. Melting ice reduces friction, which makes it hard to walk on it without falling.

fulcrum   the supporting object around which a lever (a simple machine) pivots.

gravity   a force of attraction between two objects.

heat sink   usually a specially shaped piece of metal used to carry heat away from an object. In electronics, heat sinks are attached to transistors and integrated circuits to keep them cooler.

hydraulics   a branch of physics that studies the laws of liquids in motion.

hypothesis   a thoughtful, reasoned guess about something, based on what is known. A hypothesis must be proven by experimentation.

incandescent bulbs   Most home lighting comes from incandescent or fluorescent bulbs turning electrical energy into light energy. In an incandescent bulb, electricity passes through a small wire, called a filament, which glows brightly. In a fluorescent bulb (a long, straight, or circular tube), the inner surface is coated with materials called phosphors and the tube is filled with a gas. When electricity passes through the bulb's heating element, the gas gives off rays that cause the phosphors to fluoresce (glow).

kinetic energy   the energy of work being done; the energy of motion. A bowling ball rolling down an alley is an example of kinetic energy. See potential energy.

magnetism   a force exhibited by certain objects that attract iron.

mass   the amount of "stuff" of which an object is made. The more mass it has, the heavier it is. A ping pong ball and a golf ball are about the same size and shape, but a golf ball has more mass.

microwaves   Microwaves are a kind of radio frequency energy. They are electromagnetic waves. Their frequency (the number of times the wave vibrates each second) is much higher than most other types of radio and TV waves. Microwaves are used for telephone and satellite communications as well as for cooking in "microwave ovens."

observation   using your senses—smelling, touching, looking, listening, and tasting— to study something closely, sometimes over a long period of time.

pendulum   A weight hung by a wire or string tied to a fixed point (one that doesn't move) is called a pendulum. If the weight is pulled to one side and then released to fall freely, it will swing back and forth. Gravity pulls it down, and then momentum keeps it moving past the "at rest" hanging point. Eventually, the weight stops swinging back and forth because air friction slows it down.

potential energy stored-up energy; the ability to do work. A rock resting high on a hilltop has potential energy; the ability to do work because of gravity. See kinetic energy.

power power is defined as the rate of doing work or energy used. Power equals work divided by time. Units of measure of power are the watt and horsepower.

quantify to measure an amount or "how much" of something

radio frequency energy electromagnetic waves used to carry TV and radio signals.

reverberation the sound of thousands of echoes washing together, each echo having a different delay. The ear doesn't hear any one particular echo, but rather a mixture of indistinguishable sounds.

simple machines tools—such as levers, inclined planes, pulleys, wedges, screws, and wheel and axles—that make it easier to do work. All complex mechanical machines are made up of simple machines. Simple machines provide a way to change a force to a distance, a distance to a force, or to change direction (a pulley on a flagpole changes the downward force on a rope to an opposite, upward, movement of a flag).

surface area the amount of outside area of an object.

tare weight The tare weight is the weight subtracted from a gross weight to allow for the weight of the container. The result gives the weight of the contents of the container or holder. If you want to know how much your cat weighs, but he won't sit still on a scale, weigh yourself holding the cat. Then weigh just yourself. Subtract your weight (the tare weight) from the weight of both you and the cat. The answer is the cat's weight.

tensile strength how strong something is, and how much tension or pressure it can take before it breaks. Steel has great tensile strength.

torque a twisting force, or the force used to rotate an object.

trajectory the path of an object as it travels through the air.

unbalanced force a push or pull on an object that is stronger in one direction than any push or pull in the opposite direction.

velocity speed with a direction of the motion assigned to it.

wavelength a wave is a form of energy travel, like rolling waves in the ocean. Sound and radio waves would look similar to ocean waves, if we could see them. A wave has a "crest" or peak (highest part) and a trough (lowest part). The length from crest to crest or trough to trough is the "wavelength." The wavelength of a tsunami (a tidal wave) can be 100 miles (161 kilometers) long!

weight the force of gravity pulling downward on an object, toward the Earth.

work the measure of the motion-producing effects of a force. The formula in physics for work is WORK = FORCE × DISTANCE.

Index

About the Authors

Bob Bonnet, who holds a masters degree in environmental education, has been teaching science at the junior high school level for over 25 years. He was a state naturalist at Belleplain State Forest in New Jersey. Mr. Bonnet has organized and judged many science fairs at both the local and regional levels. He has served as the chairman of the science curriculum committee for the Dennis Township School System and is a "Science Teaching Fellow" at Rowan University in New Jersey. Mr. Bonnet is listed in "Who's Who among America's Teachers."

Dan Keen holds an associate in science degree, majoring in electronic technology. Mr. Keen is the publisher of a county newspaper in southern New Jersey. He was employed in the field of electronics for 23 years and his work included electronic servicing as well as computer consulting and programming. Mr. Keen has written numerous articles for many computer magazines and trade journals since 1979. He is the coauthor of several computer programming books. For ten years he taught computer courses for adults in four schools. In 1986 and 1987 he taught computer science at Stockton State College in New Jersey.

Together Mr. Bonnet and Mr. Keen have had many articles and books published on a variety of science topics. They are the authors of the following books: *Science Fair Projects: The Environment*, *Science Fair Projects: Electricity and Electronics*, *Science Fair Projects: Flight, Space and Astronomy*, and *Science Fair Projects: Energy*, published by Sterling Publishing Company.